# DEEPER REALITIES OF EXISTENCE VOL. 1

*This Book is Presented to*

*by*

Sign

*Comments*

Date:   Time:

# DEEPER REALITIES OF EXISTENCE

-IYKE NATHAN UZORMA

Copyright © 2012 by Iyke Nathan Uzorma.

ISBN:     Softcover        978-1-4797-0800-0
          Ebook            978-1-4797-0801-7

All rights reserved. No part of this book may be reproduced or transmitted in any form or by any means, electronic or mechanical, including photocopying, recording, or by any information storage and retrieval system, without permission in writing from the copyright owner.

Unless otherwise indicated, all Scripture quotations are taken from the King James Version of the Bible.

This book was printed in the United States of America.

Published under permission in U.S.A. By:
CHRIST RESTORATION CENTRE, TULSA, USA
(APOSTLE/PROPHET MOSES AYUKETA)
+1-918-849-8130, +1-918-606-9973, brmozesa@yahoo.com, www.christrestoration.org

**To order additional copies of this book, contact:**
Xlibris Corporation
1-888-795-4274
www.Xlibris.com
Orders@Xlibris.com
121860

# DEDICATION

I Dedicate This Book To
His Excellency
OWELLE ROCHAS OKOROCHA
Governor Of Imo State, Nigeria

For alleviating the suffering of the down trodden, providing 'free education' up to the university level in Imo state, including his many years of selfless service of love to humanity, which also encapsulates his humanitarian system of governance; and to all those who work for the sanity of human race through selfless actions, words, or thoughts of love.

## CLARIFICATION OF IDENTITY
(Iyke Nathan Uzorma is not Protus Nathan Uzorma)

Following the column of Prof. Protus Nathan Uzorma (Nathan Uzorma Protus) on 'Philosophical Reflections' every Wednesday in the 'DAILY SUN' newspaper (Nigeria), several people at one time or the other, have genuinely sought to know from us whether Prof. **Iyke Nathan Uzorma, Harbinger of the Last Covenant**, (of the former Occult Grand Master, now in Christ fame) is the one writing 'Philosophical Reflections' in the 'DAILY SUN.'

Some people, however, having mistaken Protus Nathan Uzorma to be Iyke Nathan Uzorma, have cast aspersions on the latter based on the writings of the former. For instance, one Onyero Mgbejum in an article posted in the website (www.sunnewsonline.com., 'The question of virgin birth and the deity of Jesus,' Thursday April 16, 2009), judged and condemned the **Harbinger of the Last Covenant (IYKE NATHAN UZORMA)** while reacting to the article of Protus Nathan Uzorma. Part of what Onyero Mgbejum wrote, reads:

> "I read with keen interest and inclination towards asking searching questions, the articles written in your widely read Daily Sun newspaper of Wednesday, March 18, 2009. In the two articles written by Sina Adedipe

and Prof. Nathan Uzorma Protus, the questions of Virgin Birth and Deity of Our Lord and Saviour Jesus Christ are subjects of comment by the two writers. In his article Prof. Uzorma said that 'the great question to the Christian world is: Why is the virgin birth of other star beings and sages said to be demonic and that of Christ divine'.... Things of the Spirit can only be perceived by those who have the Spirit of God, those who have accepted the Lordship of the Lord Jesus Christ. People like Prof. Iyke Nathan Uzorma... is deceptive and thus should be regarded as an antichrist... Prof. Iyke should also be regarded as an atheist empowered by Satan to confuse and deceive people."

While we pray the LORD to forgive Onyero Mgbejum for this malicious error, including those who falsely allege that Iyke Nathan Uzorma has gone back to the occult and for whatever the powers of darkness have used same to achieve, we also advice anyone who claims to 'defend' the LORD to be very watchful through the Holy Spirit, so that one will not degrade the true work of Christ while 'fighting' for the Almighty. And for the avoidance of doubt, let it be known that Prof. Iyke Nathan Uzorma is the elder brother of Prof. Protus

Nathan Uzorma. Both are different persons from the same parents. While 'Nathan' is the family middle name, 'Uzorma' is the family surname.

Remain blessed!

Signed:

MRS. EME EKANEM ETUK, M.A.
Special Assistant, Outreach Affairs, To The
HARBINGER OF THE LAST COVENANT

IYKE NATHAN UZORMA

PROTUS NATHAN UZORMA

# COMMENTS ON THE MISSION AND PUBLICATIONS OF IYKE NATHAN UZORMA

"It is indeed a miracle that Prof. Iyke Nathan Uzorma, a well known Guru and Perfect Master of Esoteric Mysteries...has been humbled and convicted by the Holy Spirit to acknowledge and accept the sovereignty of the true Christ over his life."
-PROF. OLADEJO OKEDIJI
Former Dean, Faculty of Social Sciences,
University of Lagos, Nigeria.

"Though several publications on religious and spiritual matters exist, but from my personal experience, none the world over is as authoritative as the works of Prof. Iyke Nathan Uzorma, Harbinger of the Last Covenant."
-HIS EXCELLENCY,
YURI KWAKU BAAWINE
Former Ambassador of Ghana in
Saudi Arabia.

"My meeting Prof. Iyke Nathan Uzorma, Harbinger of the Last Covenant, in Nigeria, greatly lifted my spirit and opened my eyes more on a deeper aspect of the mysteries of life."
-PROF. YOSIAH MAGEMBE BWATWA
Dean, Faculty of Education, Tumaini University, Dar-es-Salaam, Tanzania.

"We thank God for using in our day and age His Harbinger, Prof. Iyke, to open the eyes of many in his spiritual mission that is divinely bound to move the world"
-CHIEF DONALD UGBAJA, NPM, mni
Deputy Inspector General of Police (rtd.), Abuja, Nigeria

"The Harbinger of the Last Covenant, who was former Occult Grand Master, appears plain and ordinary but, as revealed to me, he is a head lion in the pride of God's lions."
-PROPHETESS NNEKA A. UDI-KEN
Founder, Tabitha (Mercy) Prayer Ministry, Effurun, Warri, Nigeria.

"My respected mentor, Prof. (Apostle) Iyke Nathan Uzorma, Harbinger of the Last Covenant, thanks for teaching me the reality of the forces of darkness, their manipulations and weaknesses. Thank you sir and may God continue to use you to bless our generation."
-MOSES AYEKETA
Presiding Bishop, House of Prayer,
Christ Restoration Centre, Tulsa, Ok., USA.

"Dear Iyke Nathan Uzorma, after reading your books, I experienced a fresh and new anointing upon my life."
-PASTOR HAYFORD A. B.
Assemblies of God Church
P.O. Box 95422, Tel Aviv, Israel.

"Your books completely changed my weakness in spirit. To be precise, I was a Muslim and converted into the Christian Religion and got salvation through Christ."
-PRINCE HABIB WINFUL
Harter Str. 69 Graz-8053
Austria, West Europe.

"Man of God, I thank God for your book which helped to strengthen me in the Power of Jesus Christ."
-NATHAN C. U.
26 Friedrich Ebert Street,
57-518 Betzdorf, Germany.

"I read your book and the blessing I got from it is beyond what I can express.
-DAVIDSON O.
Don Whari 445-2 Kyonggid 0445-890
Seoul, South Korea.

"I must say that your book is an overall experience compared to others whose testimonies we've read."
-REV'D. C.S.D. OSUIGWE
Onitsha District H.Q.
The Apostolic Church of Nigeria.

"Your book has strengthened me in my Christain life. I am now anxious to read other books by you."
-RICHARD REMNARINE
Plot 8, Springlands Corriverton Berbice,
Guyana, South America.

"Words are inadequate for me to express the contents of your books. I thank God for the enlightenment you have given us."
-REV'D. ROBERT ARYIKEN
29, Rosemead Ave,, Mitcham Surrey
CR4 IEZ England, United Kingdom.

"I came across your book and could not put it aside but had to finish it in hours. I thank God so much for your life."
-PROF. P.A.DUAH
University of Ghana, Legon, Accra.

"Your books have changed my life and the lives of friends and families."
-W.P. LIESCHING
P.O.Box 2334, White River 1240, South Africa.

"The books of Prof. Iyke Uzorma, Harbinger of the Last Covenant, are nobel and highly enriching to humanity."
-PROF. SMART O. NWAOKORO
Head, Dept. Of Animal Science,
University of Benin, Nigeria

# CONTENTS

Dedication......................................................v
Clarification of Identity...............................vi
Comments....................................................ix
Acknowledgments........................................xvii
Prelude.........................................................xix
Introduction................................................xxvi

## CHAPTERS:

1. Echoes Of Awakening..............................32
   *What Leads You  * Know Thyself
   *The Lower Path of Life
   *The Higher Path of Life

2. The Core Of Earth Winnowing..............88
   *Basis of Earth Winnowing
   *Transforming The World
   *The Living Earth
   *Gross Material Limitation
   *Cosmological Verdict
   *Balancing Unit of Dual Powers

3. Planetary Mystery...................................48
   *Ascent in The Solar System
   *Relativity of Creature  * Hidden Things
   *The Base of Manifestation
   *Evidence of Earth's Consciousness
   *Earth Can Help You

4. Summary Of The Mystery Of Life.........59
    * Rest in Peace ? * Reality of Mental Act
    * Dichotomy of Inner Pattern
    * Cosmic Expression of Oneness
    * Light in Motion * Radiation of Power
    * Extraterrestrial Intervention
    *Salvaging Humanity * Know More

5. Most Vicious Astral Manipulation.........70
    *Burden of a Planet
    *Little Creatures, Man and Earth
    *The Invaded Mind
    *Symbol of Hate
    *Mundane Limitation
    * Facing The Reality

6. Dimensions Of Existence........................81
    *Controller of Manifestations
    *Other Dimensions
    *Influence of Mind Over Matter
    *A strange Fact
    *Vicious Thought Projections
    *Understanding The Relativity of Nature

7. The Game That Saved............................92
    *Surviving on Humans
    *Signet of Mystery
    *Astral Right  *Cosmic Crime
    *Basis of Psychic Attack
    *Man's Prime Duty
    *The Game of Jesus the Christ
    * Savior of the World

8. The Eternal Validity Of Oneness............103
   *Higher Path of Life
   *Essence of Religion
   *Transcending the Mundane Religion
   *The Journey Towards God
   *The Only Way to God
   *Masters of Darkness
   *The Point That Controls Everything
   *The Ultimate Source
   *The Best Path of Victory
   *The Age of Winnowing
   *Conflict of Powers
   *Man Against Man
   *Childish Nature of Earthmen
   *Akashic Records *Messengers of Light

9. The Reality Of Our Time........................127
   *Nothing Happens by Chance
   *The Biblical Lot
   *Now is The Time to Act
   *Fighting With The Wrong Weapon
   *Lovers of Humanity
   *Escalating Injustice With Injustice
   *Willing Tools of Darkness
   *Mundane Religious Propensity
   *The Only True Religion
   *The Path of Light
   *Seal of the Christ
   *Mystery of Our Time
   *War is On

10. The Battles To Sustain Our World........148
    *Gaining Grounds of Victory
    *Regrouping of Forces
    *My Personal Experience
    *Winnowing of Earth
    *Manifestations of God
    *Supreme Plan to Salvage Earth

11. Framework For A New World..............159
    *Higher Existence in Light
    *Expected Changes *The Future Man
    *Earth in The World of Stars
    * Play Your Part Now *Psychic Terrorism

12. Walking On The Vicious Path...............170
    * Thoughts Potency
    *The Making of a New Man
    *The Realm of Universal Freedom
    *Take Less Give More
    *Contributing to The Pool of Darkness
    *Hindering Your Rise in Light
    *Are You Your Own Enemy?
    *Overcome The Path of Hate

13. Potencies Of Universal Love.................183
    *Astral Intricate Web
    *The Transforming Power
    *Let There be Love
    *Universal Constitution
    *We Are One  *Evidence of Light

Harbinger Books............................................195

# ACKNOWLEDGMENTS

My beloved wife, Lady Udeme Uzorma, you are the great instrument that encouraged me in diverse ways to write this book. I sincerely appreciate all your efforts. My good friend, Koffi Panford, your love and willingness to advance the course of this book, which made you to go even extra miles, will ever remain indelible in my heart.

My special thanks to a great woman, my beloved senior daughter and mama, Anthonia Nneka Udi-Ken (Supreme Prophetess), spiritual leader and founder of Tabitha Mercy ministry, who greatly encouraged the publication of this book and who with her household and spiritual children stand at all times to uphold the divine mission of the 'Harbinger of the Last Covenant'.

I am indebted to my editorial team of the Harbinger Digest, especially my big brother Kenneth Udi, my big daughter Eme Etuk, Uche Melisa, Hilary Okolie and Amos Livingstone, for their editorial assistance; so also are A.J. Okoro and Ntekim Rex for their organizing assistance.

My lovely daughter Pat was splendid in her numerous constructive questions on the deeper realities and mysteries of life, which greatly quickened the expositions in this book whilst at the formation of the manuscript. You are truly appreciated. My Personal Assistant Michael Isemin and my Private Secretary Chibuzo did well in coordinating the publication of this book, including Churkylove for the cover design, which I greatly appreciate.

God bless you all!

-Iyke Nathan Uzorma
Thursday 8 March 2012

# PRELUDE

Deeper Realities of Existence, is a book that holds for mankind a message of profound truth and revelation of hidden mysteries. It is the panacea to all forms of physical, psychological and psychic terrorism, including vicious Astral attack. This book elucidates the basis of Planetary Winnowing in this Age, the core terrestrial and extraterrestrial danger of atomic radiation unknown to mundane scientists, Universal Signet of the Supreme Mastership of Christ, Immutable Laws of the Universe, the rise of a sane civilization of Universal Brotherhood on Earth, amongst others.

When I came out of the most vicious network of darkness into the Light of the One Eternal God, I was made to believe by some 'believers' that the Almighty God spoke to the human race only in the Bible. On the strength of this belief-system, not by what I know or by what the Lord told me, I strongly condemned any other book on Earth which claims to contain what God said to humans.

Thereafter, I was bamboozled when the Lord manifested and told me that there are certain ancient and contemporary sacred writings that also contain 'His Words' to humanity at diverse times and places, in different parts of the world. Following this, I made

Deeper Realities...

research in the Bible and found that the Holy Scriptures never at any time claimed that the Lord of the Universe only appeared to the Jewish Prophets, or spoke to the children of men only through the Prophets from ancient Israel.

To the contrary, the foremost evidence that God also spoke through some humans in places other than Israel, is found in the Holy Bible itself. I think it is not proper to condemn what you don't know, rather we should do so on the strength of what we know. Saint Paul admonished that we should "Prove all things; hold fast that which is good" (1Thess. 5: 21).

You should first study something before you can prove it true or false, otherwise you don't have the basis to condemn it. Thus, when I took time to study the Koran, Divine Iliad, Bhagavad-Gita, amongst others, I did so with the intention to know what the Lord said in them, to prove such for myself while holding fast to that which is good. There is nothing good beyond Divine Love, which is the Lord Himself. As such, some of the things spoken by the Lord or His Divine Ones above in this connection in diverse places, are brought into this work, for the guidance of the entire human race.

Take Abimelech, for instance; he was a Philistine king, but God still appeared and spoke to him in a dream (See Gen. 20:1-8). Is this not a clue to the fact that God did appear and spoke to diverse people in any matter He wished? If God spoke to one who was not a Jew, whilst a publication came forth of what the Lord of Creation said, should we study or discard same? Should we discard a thing said by the Lord, simply because the human channel He used did not come from the Jewish stock?

What about Cyrus? He was the king who led Persia to conquer Babylon in 539 BC. To him the Lord said: "He is my shepherd, and shall perform all my pleasure" (Isa.44:28). Is this not a clue that in the history of mankind, there were men like 'Cyrus to the Lord' in different parts of the world? Certainly, there were men used by God in different parts of the world on the part of Light as 'shepherds' for the good of men.

Furthermore, on the Lord speaking to and through some human channels other than the Jewish Prophets, we should also note that even in the days of the Great Prophet Moses, the Lord spoke through a man called Balaam. Some Bible Scholars hold that Balaam was a soothsayer. For instance, the 'Bridge Bible Directory' (an A to Z of Biblical information) page 40 called him "the soothsayer Balaam". The 'Insight' of the 'New International Version' of the Bible (NIV) page 136, has it that "Balaam was evidently a professional magician". However, the 'Drake's Annotated Reference Bible', King James Version, page 10 of the Complete ConcordanceCyclopedic Index, holds that Balaam was a "True Prophet before backsliding".

**Deeper Realities...**

Whatever he was, however, the fact remains that the Lord appeared to him, spoke to him and spoke through him. Balaam, the son of Beor, hailed from Pethor in Mesopotamia, and as such was not a Jew. Nevertheless, we read in the Holy Bible that:

Balaam told those sent to consult him by king Balak of Moab against Israel that "I will bring you word ... as the LORD shall speak unto me" (Num. 22:8). Meaning that the LORD has been speaking to him before then, on which he relied that He would still speak.

The Lord manifested and visited Balaam: "And God came unto Balaam and said..." (Num. 22:9). "And God came unto Balaam at night" (Num. 22:20).

Balaam worked for "the rewards of divination" (Num. 22:7). The Angel of the Lord at a point said to him: "thy way is perverse before me" (Num. 22:32). Nevertheless, "the Spirit God came upon him" (Num. 24:2) and he was directed to say only the things God gave him, which he did, (See Num. 22:35), even as "God met Balaam" (Num. 23:4) and "put a word in Balaam's mouth" (Num. 23:5).

To me, if the Lord puts his words in anyone's mouth, past or present, good or bad, in any part of the world, and such words from the Lord are published, I would gladly peruse it. For what matters to me in this connection are the words of the Lord and not the human channel He used. To discard a book containing the sayings of God simply because the human channel used to express same does not conform to the system of my religion or my church, is childish in a higher spiritual sense. Saint Paul wrote:

> "When I was a child, I spake as a child, I understood as a child; I thought as a child: but when I became a man, I put away childish things" (1 Cor. 13:11).

The foregoing is for the guidance of the reader of this book. As you will observe, references are made in this book not only from the words of God as recorded in the Bible, but also from the words of God as recorded in other sacred writings, past and present. This is for the overall good and elevation of the entire human race in Light, not merely the elevation of a particular religion, tribe, church, or nation. My aim here is for this to culminate in the consciousness of the brotherhood of all existence within the framework of higher realities.

Also, it is important to note that when mention is made in this treatise of 'High Metaphysics', this has nothing to do with contemporary metaphysics or mundane philosophy. Of course metaphysics deals with the abstract studies of the nature of reality; as a branch of philosophy, it deals with the initial principle of things, which includes ontology and cosmology. Noetic science is also a fundamental aspect of metaphysics. All subjective or abstract matters beyond the verification of the five senses and all mundane data, are metaphysical.

Thus, when in strict religious terms we speak of our prayers going to God, the eternal existence of Soul, Angels of God in other realms of life, life after death, amongst others, we bring religion into the platform of metaphysics, abstract studies. Consequently, religion and metaphysics must interwove and interchange in diverse angles.

Condensed metaphysics which we study in schools, when opposed to what the Lord or His Angels say, must be considered feeble, rising only from the mundane limitation of the human mind and not from the Divine. Thus, this in many respects could mislead, especially when the terrestrial recognition of metaphysics comes purely by speculation.

However, the knowledge of a thing for the good of humanity, though abstract, but based on a bona-fide information from Extraterrestrial Intelligences in the higher realms of Existence, is 'High Metaphysics', which can also be called 'Spiritual' or 'Divine'. Now, Extraterrestrial Intelligences in the heavenly realms includes the manifestations of God or the Lord in such realms, as well as the Angels, Archangels, Watchers and other Great Ones within the systems of the 'Forces of Light'

Such Divine Intelligence live constantly on a higher, possibly the highest, gestalt of consciousness, within the frequency of the highest Universal Love, Power and Divine Knowledge. They operate far beyond the earthman's organized religions, metaphysics, philosophies and occult fraternities. For now, they represent in certain aspects the height which man will attain someday in the endless journey of life.

Quite clearly, they are beyond the earthly, the mundane, or the terrestrial system of reality. Having gone beyond the terrestrial framework of awareness, they are therefore rightly called 'Extraterrestrials', regardless of whatever other names they may be called. The knowledge, High Metaphysics, which they bring for the guidance of the entire human race, flows from the Eternal Essence of the Holy Spirit. Please note these as you read this book.

To this end, 'High Metaphysics' is used in this book to connote any abstract information from the Lord or His Angels to humans. Also take note of this, as I present this work to you beyond the limitation of our organized religions on Earth. And if you have any question arising from this book, send it to me with your name, city and Country, so that it would be answered in the second volume of this treatise.

Peace be with you.
-The Author.

Deeper Realities...

## INTRODUCTION

Under the sway of wrong thinking and vicious mental pattern, the earthman is bereft of the consciousness of Divine Light and the potencies manifested within the system of selfless Love. Thus, he parades the ignorance of his core multidimensional eternal identity within the mundane camouflage. In this camouflage system, evident in the physical realm of Earth, humans become subjugated by the illusory propensities from which hate flows. Consequently, man is forced by the influence of the mode of hate to submit his life to diverse perverted thoughts, words, or actions that lower the human existence in the Cosmic Scale of the One Eternal Spirit, GOD.

The prime duty of man in the world of man includes, among other things, to recognize and live within the framework of the higher path of life via his Higher Nature, his essential Divine Aspect. This Divine Aspect of man arose from the Primordial Specie of THAT which we call GOD. It is the Higher Self by which one can make conscious contact with his Source in diverse versions and with all that lie beyond the mundane. The establishment of this contact is brought about within the ambit and manifestations of the consummate essence of Universal Love.

The foregoing is imperative, because any activity within the realm of Earth, or within any realm of the Cosmic manifestations, which does not dovetail with the consciousness of Love transcending the temporal system of vicious life, cannot foster sane existence as well as the elevation of anyone in the journey of life.

The earthman in his Higher Aspect is a Citizen of the Universe. If he understands, upholds and manifests this recognition within the moment point of awareness, he remains a bona-fide citizen, not a slave, whilst in the world of man. In the realm of the Light of Love, man is a Universal Citizen; in the realm of the darkness of hate and vicious mental pattern, man becomes a slave, a captive, in the physical system of reality. Here the vicious Astral forces come into play with diverse forms of psychic, physical and psychological terrorism, amongst others, often via human channels, all which are explained in this treatise.

No man is free on the physical Earth from the multifarious web of the vicious Astral forces, until he comes consciously to the platform of life in Light, the higher path of life, which Love sets in motion. Then he will begin to interact with the Forces of Light and interchange the essence of the deeper realities of existence in the higher frequencies of GOD, both here on Earth and in the upper (heavenly) Planetary Systems of the Cosmos. Consequently, the earthman will join the Great Ones in the Universe who work for the upliftment

**Deeper Realities...**

of the human race, via the Cosmic Law of Homogeneity. In so doing, man will be brought face to face with the deeper realities of what Planetary and Inter-planetary existence is all about, the existence which constitute the validity of his entire life that he should know. These are also explained in this book.

When the earthman awakes from the mundane life, he begins to explore the characteristics of the '**Kingdom of God within himself**'. The dimension of universal spiritual cum metaphysical experience that comes to him in this regard becomes eternally valid, for it goes beyond the limit imposed on man via the psychology and belief systems of organized religions, which in some cases give birth to the most vicious religious ideologies.

We must understand, however, that the process of reading the sacred writings of world religions, or hearing about God from others, is valid to the extent that it brings the Supreme Light into the formation of inner experience, but feeble in approximation to the inner experience gained within the framework of this Eternal Light. Your core identity, the Spirit which you are now, not which you will become after death in human terms, is quickened to accelerate and culminate the inner experience of Being into objective embellishment, for the sanity of individual and collective existence of humans, by the elements of Universal Love.

This experience is not impossible but possible for

every concentrated individualized Spirit (Soul personality) in the human embodiment, every living man. This book is sure to be a transgression on my part if its essence holds that which is impossible for the earthman to comprehend. The very essence of this treatise does not tell man what he is not capable of knowing, what he is not capable of putting into action. This work encapsulates the existence of Love on the physical realm of Earth and other realms of life. It holds that which man is able to understand and live by.

I am fully aware that it is a transgression or a mockery by the standard of Universal Laws, to tell man what he is not capable of knowing and putting into action. God spoke about this in the *Divine Iliad* thus:

> "When I commune with knowing man in Light I command him, e'en though he knowest all of Me as Me, to tell no man of Me beyond that which he can bear. I limit him, as I do thee, to the comprehension of his day and age, for man cannot bear that knowledge of Me which is too far beyond his day and age".

The Great Lord Jesus the Christ whilst in the world of man also observed this principle when He told the earthmen of His time: "I have yet many things to say unto you, but ye cannot bear them now" (John 16: 12). Those 'many things' were not said because they were far beyond the people of that time, though they remain valid things that man should know at a stage.

This book is designed to lead man into the higher path of life beyond the mere mundane. In doing this, attempt is made from my point of understanding to bring forth deeper realities of man's existence on Planet Earth. The knowledge hereinafter expressed is solely embedded in Love as the core of Earth winnowing. The reader will observe that this is the central point of this treatise. Though quotations are made here from different sacred writings of world religions, the totality of this work is not within the limitation and confines of any particular organized religion. Thus, this book is not for any particular religious sect on Earth, rather it is a book for the guidance of each and everyone in the world of man, who seeks the sanity of life.

'Deeper Realities of Existence' now in your hand did not come to you by chance. There is surely a purpose by which it has come your way. I do hope you will read this book with an open mind and be clear of the purpose of it in your life and in the world of man. It is my sincere desire that the essence of this work will sink into the consciousness of the reader.

This book is specifically made for humans to think, speak and act for the elevation of the human race within the framework of sanity. It is the panacea to all forms of physical, psychological and psychic terrorism, as well as all forms of individual and collective vicious life. To come to this platform in full awareness, we are expected to know deeper things of life both in the physical realm and other realms of existence. We think, speak and act within the limit of what we know.

He who fails to know more does not grow in the journey of life. He who fails to wake up cannot stand. This treatise encapsulates the true essence of all the religious sacred writings culminating in High Metaphysics within the dimension of the Holy Spirit of Light. It sets the stage for what everyone should know, the stage of the knowledge for a better action. This is the hour to know more!

- Prof. Iyke Nathan Uzorma
(HARBINGER OF THE LAST COVENANT)

Deeper Realities...

## CHAPTER 1

### ECHOES OF AWAKENING

Earthman Awake! You now live on the threshold of Planetary Winnowing. The Universal Law by which this comes forth does not allow anyone again to sit on the fence. The era in which you stood at the gate, you refused to go in and you stopped others from entering, is gone. You were told by the Wise Ones of ancient times never to force or drag a goat to the market. He who drags a goat to the market should remember that he will also be seen with the goat in the market.

Earthman, you are not even allowed in this Age by Law to stand on the middle path. This does not exist again. You will either stand on the path of Love and go right or you stand on the vicious path of hate and go left. The choice is yours. You will experience what belongs to the path you choose. In this Age, you will either be firm and follow the path of Light to rise, or you remain behind and sink on the scale of life.

## WHAT LEADS YOU

Earthman, know and understand, and always bear in mind, that what leads you in the journey of life determines the factor of your victory or failure, your rise or fall. I tell you, if you are led by the 'spirit of a goat', you will surely be defeated in the battles of life. If you are led by the 'spirit of a lion', you will overcome all the elements that seek to bind you in the web of the mundane. Surely, it takes a lion's heart to awake, to transcend the propensities of darkness, to rise into the higher realms of Light and to be a true man.

Earthman, I speak to he who has ears to hear: Consider a battle field in which one million lions led by a goat came against one million sheep led by a lion. It came to pass that the manifestation of victory stood, not on the side of the one million lions, but on the side of the one million sheep. The manifestation of victory followed the principles of existence, by which your victory or failure is determined by what leads you.

Earthman, lie low! Be satisfied in life and where you are placed in the world of man. Then from there rise into the realms of Light via the thoughts, words and actions of selfless Love. Do not behave like a fowl that is drunk. Lie low and consider THAT which lies above you, THAT which is beyond you. What do you think will happen when a fowl that is drunk meets with a mad wolf in a battle? Everyone can guess the outcome.

Deeper Realities...

## KNOW THYSELF

Earthman, it is said by the Wise Ones of old, 'Know Thyself'. To do this entails to know more; and to know more is to fully know who and what you are, without which you cannot know what you will become. Do you truly know in practical terms who and what you are including from whence you came? If you do not know where you are coming from, how would you know where you are going to? If you do not know your destination, how would you know when you reach there?

Earthman Awake! Many times you will fall in the journey of life, many times you will rise to continue your journey. Twice blessed is he who learns from his failure and stand to fall no more. Sevenfold blessed is he who learns from the failure of another, from the mistakes of others, and be guided from the lessons of that. Surely, failure is not good for man; the only good in it is the lesson of how not to fail again. I tell you, mistake and failure are part of the existence of the lower aspects of humans. But again I say, the only real mistake or failure in life is the one from which you learn nothing.

Earthman, rise beyond the limitations of your lower aspects and come to the platform of the Higher Aspects of your true identity. Do not be gratified within the limit set in your lower aspects, from which you are subjugated, and in which you are held bound in the mundane. Do not behave like a goat that feeds where it is tied. You should seek to have experience from the Higher Aspects of your Being. You cannot say that pepper is hot until you taste it. The hour has come for man to rise above darkness that has tied many for so long. The hour has come for man to walk on the path of Light. The vicious path of darkness has not and cannot elevate anyone. For no matter how 'hot' the anger of a man is, it can never cook his food.

Earthman, you may claim to be ignorant of what constitutes the path of Light and the path of darkness in the journey of life. But I say to you: your thoughts, words and actions hold the answer. Your way of life is like pregnancy that does not hide. Is the cost of obedience not less than the price of disobedience? Is a woman not as old as she feels whilst a man is as young as he behaves? First be sincere to yourself. Yet you may turn around to ask me 'what is darkness?' You may desire that I show you in practical terms what constitutes darkness, the lower part of your life.

## THE LOWER PATH OF LIFE

Earthman, do you not know that darkness is but the lower path of life? Darkness is nothing but the absence

**Deeper Realities...**

of Light. Darkness is the realm of hate and vicious life which harbors the forces of its kind. Darkness is the system of consciousness that upholds the mundane propensities arising from lust, greed, vanity, anger and attachment to the material things of life.

Earthman, in darkness you are but an earthworm. But by your first strong decision to walk on the higher path of life, the life of Universal Love in Light, you transcend the coarse position of earthworm and transmute into a serpent, even a strong serpent. Is a strong serpent not ahead of an earthworm in the journey of life? When you become firm and fully established on the path of Light, the higher path of life, then you will bestir that which limits the serpent to excel, to transmute into an eagle, yea, even a strong eagle. Then you will rise; then you will ascend; then you will fly away in the speed of Light.

## THE HIGHER PATH OF LIFE

Earthman, will you turn around and say to me 'Harbinger of the Last Covenant, what is Light?' Do you not know that Light is solely the absence of darkness? I tell you, Light is the primordial Life Current of the One Eternal Spirit (GOD) in seething motion, within and beyond the systems of Matter, Space, Time and Energy, vibrating in the rhythm of rapid thought. Light also includes the existence of sanity and the realm which harbors the forces of its kind. Light upholds the order of the higher path of life, of which Universal Love is its ultimate essence as well as the central coordinating point of activities within its system.

Earthman, when a 'Mighty Will' of one who has become an eagle stirs up the ether, via the consciousness embedded in the system of thought of Jesus the Christ, manifesting absolute Love and radiating inner Divine Peace and Spiritual Joy, whilst their vibrations reach the higher realms of Light, then there will be Light. Within this framework, Light will come forth and manifest in the life of man and his environment.

O earthman, he who comes to this point of consciousness will see the finger of darkness, its activities and forces, no more.

Deeper Realities...

## CHAPTER 2

### THE CORE OF EARTH WINNOWING

This book is born out of a sincere appreciation for the numerous readers of my expositions who are yearning to have me speak on deeper realities of life. My aim here is for man to know more and come to the higher systems of reality transcending the gestalt of illusion and darkness, beyond the mere mundane, beyond the cycle of vicious life.

For the earthman to rise beyond the illusion of mundane existence, he must in practical terms live within the framework of the message of Universal Love. The essence of this is evident in diverse degrees in all the great sacred writings given to man past and present, such as the Bible, Koran, Bhagavad-Gita, Upanishads, Divine Iliad, The Everlasting Gospel, amongst others, of Which to me the Holy Bible is foremost.

## BASIS OF EARTH WINNOWING

The practice of Love is the true wisdom of Supreme Light, the true essence of existence and evidence of true religion. It is also the basis of winnowing of Earth and the entire Cosmic manifestations. It flows into the world of man from the 'Heart of Universal Love' solely for the elevation of the human race.

In our human ignorance cum mundane limitation, we generate hate and use same to fight for things of material value. We fight for our temples, churches, mosques, etc. We know not that the true temple, church, or mosque is the practice of Love, which is Light, the nectar of Universal Life, the life filled with the 'One Eternal Spirit' we call 'GOD'.

The manifestation of valid existence, in the consciousness of the brotherhood of life, whether in a family, community, nation, or the entire world of man, can never rise from the realm of hate. All forms of hatred in all ramifications are nothing but darkness, the forces of darkness, in motion. It is like a dark glass that covers the vision of life.

The Divine Messages sent to man from the higher realms of Light, past and present, hold that the basis of valid existence is not rancor, the manifestation of hate, for this scatters the fruit of life. It is also not vicious segregation arising from hate, which lowers the human life in the scale of the Almighty. This alone is Love, the eternal immutable principle set forth to lift man and every creature on the way to God, on the infinite path of the realization of God's Multidimensional Potencies.

## TRANSFORMING THE WORLD

It follows that LOVE seeks to guide man, and man will surely receive this guidance and be elevated only when he lives in Love. To revive or revitalize the earthman, and bring him to the recognition of the most important thing he needs, the message of Divine Love flows into the world of man from time to time, again and again, via diverse human channels. The most important thing which man lacks is the expression of Universal Love in the recognition of the unity of all manifestations. Man is truly not in lack of money, house, wife, children and all the paraphernalia of aggrandizement of the physical system of existence; the most important thing he lacks is Love, Divine Love.

The man who seeks to be quickened by the One Eternal Spirit (that is to have specific touch of the Holy Spirit, or revival in religious terms) will surely fail if he has no iota of Love. He may experience an emotional or psychic phantasmagoria and hold same to be what he seeks. It is like a man seeking for gold but does not know how it looks like; he receives a shining stone and considers same to be what he seeks.

The sacred practice of Universal Love is the final essence that will winnow and transform the world of man. And in this generation of men, the hour of fulfillment in which we live, (the Age of Divine Civilization and the Brotherhood of Life, which some call the 'Aquarian Age'), Love must, I say must, be the sole basis of existence at last on Earth.

## THE LIVING EARTH

Earthman, have you ever considered the position of Planet Earth, the world wherein we live? Do you think that our Planet is a mere mass of gross matter floating in the space, as held by some mundane scientists? Do we truly know what Earth is? Do we truly know what Planets are? Do we truly know the high degree of Love, Universal Love, expressed by Planet Earth in harboring the children of men to her limitation in certain aspects? The knowledge of these amongst others is needed for man to rise in the journey of life.

Now, earthman, do you know that like yourself, like humans, yet far more than that with millions of years

Deeper Realities...

ahead of humans, our Earth Planet is certainly a living conscious Being? As a full conscious Being, Earth Planet has her consummate dimension of existence completely outside the framework of the awareness, recognition and knowledge of humanity.

The Love we show on Earth, whether within or outside the religious system, as well as the hate we manifest, have specific impression on the subtle layers and even on the core of the Being we know as Earth - the world wherein we live. In this connection, the 'Universal Immutable Law Of Homogeneity' plays out. I hope to make this clear in the subsequent parts of this treatise.

## GROSS MATERIAL LIMITATION

However, it is evident that the human mind as a rule does not comprehend the multitudinous gestalt of the consciousness of a Planet, including our Earth, it holds via a tiny mental speculation that Earth is devoid of the awareness of her existence. It also holds the fallacy that other Planetary Systems do not harbor living Beings; for it can not comprehend on the mundane level the conscious existence of Beings not sustained by oxygen, the kind of existence we have on Earth.

Consequently, and by virtue of the limitations inherent in the physical system of existence, man does not recognize the dimension of Universal Love that a Planet can show to embellish Inter-planetary existence in the 'Universal Body' of the Almighty God. He may not know and may fail to cooperate with Earth on the conscious magnitude of Love she showers on humans for our ascent in the Solar System towards 'God's Greater Light'.

Every creature, whether big or small, has consciousness (awareness unit of being) within the system of a framework in which it is embedded. The size of a form manifested in creation is in the first place evidence of primordial consciousness on a certain degree of intensity. Thus, the diverse species or forms of manifestation in the macrocosm, solely reflects the variegated intensities of consciousness. For consciousness is the basis of certainty and manifestation of a form within the entire Universal Systems. It cannot be otherwise.

All forms of manifestation therefore, including Planets, humans, ants, amongst others, have life, from which a particular kind of consciousness springs. The Supreme Absolute Spirit (GOD) is the Fountainhead of Life-Currents. HE is the Life-Current Itself. The Great Lord Jesus the Christ, the anthropomorphous cum humanoid manifestation of God in the world of man, said to the earthmen: "I am the resurrection, and the life" (John 11:25).

Deeper Realities...

## COSMOLOGICAL VERDICT

In the beginning, (not the beginning of GOD, for GOD HIMSELF is both the beginningless as well as the base before the beginning), the Supreme Absolute Spirit placed a minute portion of HIS Version we know as Life-Current, Life-Energy, or the Spirit of God, in the Creative Matrix of HIS Powers, out of which the entire Creation was born in diverse 'family' forms.

Our Earth is a member of the family of Planets, which includes the Stars, Moons, Suns, etc. The earthman is a member of the family of humanity. Man is a living conscious Being of a kind, while Earth is a living conscious Being of another kind. Birds, microbes, the mighty Angelic Beings and a host of other life forms in the entire Cosmic manifestations, have their family units on a certain kind of awareness.

The denizens of the 'human family' are not more in number than the 'Planetary Family' in which our Earth belongs. For instance, the population of humans on Earth has not reached 8 billion. The number of Stars in the entire Cosmic manifestations are not totally known. According to astronomers, the human eye can, on a clear night, see about 4,500 Stars, while the strength of a small observatory telescope aids us to see about 2 million Stars. Today, thousands of millions of Stars can be seen via powerful modern telescopes in our Milky Way

Galaxy (the system of Stars that contains our Sun and his Planets) alone.

Astronomers hold that there is a cluster of Milky Ways, some far larger than our Milky Way, of between 18 to 20 Galaxies, which contain several billions of Stars apart from Planets. Yet they hold this vast number to be less when compared with the 'Spiral Nebulae' system in the infinite space.

However, the cosmological verdict of High Metaphysics, based on information coming to man from Extraterrestrial Intelligences in the Higher Planetary Systems, holds that there are not between 18 to 20 but nine hundred million Galactic Systems. In each of these is a Sun with his Planets, Moons, Spiral Nebulae and unnumbered Stars. This is the 'family' where our Earth belongs; it far out-numbers the children of men in this physical realm of existence.

Our Earth when considered in the light of the 'family' to which it belongs, becomes a minute portion of same, just as each man is a minute portion in the realm of humanity. Some Planets even in our Solar System are far bigger than the Earth. For instance, Jupiter is above 500 times bigger than the Earth. An air-craft on the speed of 600km per hour will go round the Earth within 41 hours. But on the same speed, it will go round the Sun within 4,368 hours, six months. Again, there is a Central Sun, the size which is beyond mundane speculation. The Central Sun has its order of existence on 95 billion miles above the order of the nine hundred million Galactic Systems, around which all the Galactic Systems travel at the rate of 15,004 miles per second, returning to the point of departure in about 26,000 years.

## BALANCING UNIT OF DUAL POWERS

All systems of existence, all forms of manifestation, are governed by Inexorable or Immutable Cosmic Laws proclaimed as the 'Will of God' in religious terms. The Law of Balance is one of these, for it interchanges the polarized cum dual Powers of the Almighty manifested, (the Positive and Negative Currents of God in the realm of physical, metaphysical and spiritual existence), in all poles of opposite representation. This Law records the 'Oneness in GOD' of all dualities and divisions of seemingly separate multifarious parts of existence.

The expression of God's systems of 'positive and negative currents' as well as HIS duality is evident in all the sacred writings. Thus, in the Bible God said:

> "I form the light, and create darkness: I make peace, and create evil: I the LORD do all these things"
> (Isa. 45:7)

## The Core of Earth Winnowing

In the Divine Iliad, God said:

> "Know thou then that I alone live. I do not die but out of ME comes both seeming life and death. Know thou also that the divisions of MY thinking are but equal halves of one. For I again say that I AM One; all that comes from ME are One, but divided to appear as two."

Now, the human form of existence is the foremost balancing unit on Earth of the dual currents of power. Consequently, man is capable of thinking positive or negative and manifesting Love or hate. Man himself is a highly concentrated unit of the Spirit of God that holds together in his Higher Aspect diverse frequencies of Light, which Love sets in motion. He also holds in his lower aspect certain propensities opposed to Light, which hate sets in motion. Both are balanced via a magnetic force of the mind within certain degree of applied mental compression. The human life is programmed to rise into higher levels of God's Primordial Light through the expression of Love, or be lowered in the stratum of existence, via the expression of hate; all which have marks on our Earth Planet through the 'Law of Homogeneity'.

Deeper Realities...

## CHAPTER 3

### PLANETARY MYSTERY

The One Eternal Omnipresent Spirit (GOD) is ultimately the Absolute Supreme Light that man can experience and express within the framework of Universal Love, Divine Love. Love is God's Light in motion, the essence of a higher existence. Within the Light of the Universal Body of the Almighty, there is neither religion nor philosophy; what exists is only the 'Consummate Cosmic Stream of Love.'

Some mighty Beings in the higher frequencies and realms of life are established in Light. They operate as what we call the 'Forces of Light'. Some mighty Beings also, having acquired the current opposed to Light, are enamored in the mephistophelian nature and bereft of the consciousness of good. Their pleasure lie in the pain of others. These are the proclaimed 'forces of darkness'. The multifarious battles between the Forces of Light and the forces of darkness in our Solar System, including the Earth, plus the role of man in it, are matters clarified in this book.

## ASCENT IN THE SOLAR SYSTEM

For now we should strive to comprehend in practical terms the fundamental role that Light which is Love, plays in our individual and collective ascent in the strata of Planetary and Inter-planetary existence. The 'Stream of Love' is the Central Power and Light of the Absolute Spirit. In all the realms of Terrestrial, Extraterrestrial and Spiritual life, within and beyond the Cosmic manifestations, the 'Stream of Love' is the only true philosophy, religion or metaphysics. It is also the bona-fide giver of these. It alone binds all things for ascent in eternity to the Supreme Source of Life.

The practical realization of this constitutes in certain ways the Divine embellishment of the Earth Planet, for the present and future man. This knowledge is made not to replace but enhance the profound understanding of all that God said to humans in the sacred writings via some great human channels of Light.

We should try to understand why such human channels of the past, (lets leave the present for now), whose lives were considered extra-ordinary in their time, proclaimed the message of Divine Love on diverse degrees. The teachings of Lord Buddha, Guru Nanak, the Holy Prophet Muhammad, Lord Shri Krishna, the Great Prophet Moses, Socrates, Lao-tzu, Confucius, amongst others, have inherent in them principles and elements of Love.

*Deeper Realities...*

The Savior of the World, Our Lord Jesus the Christ, went further to demonstrate what practical Universal Love is. He set specific standard which any earthman can follow (not merely within the confines of organized religion) to be established in the Light of God. In my understanding, this is based on the true recognition of the high impression of Light that the practice of Love makes in all forms of manifestation and beyond. Love stimulates benevolent impression on the human subtle faculties, including the gross material aspect, as well as on the core of the great Being we know as Earth. It accelerates our journey on the way to God, both in this present life and the life beyond death.

## RELATIVITY OF CREATURES

Now, we live as little creatures on the mighty body of a Planet, without fully knowing what the Earth is in her consummate core subtle nature. Also other lesser creatures live on us without knowing what we humans are, yet they live in a world which is man. Minute creatures like the parasitic worm in the human intestines live in a 'world' which is the human body. Creatures such as lice live on man's unkempt hair. These little creatures and others do not know the true nature of even the physical aspect of man. They have their own framework of existence and can never understand what man is from

Planetary Mystery

man's point of view. Our days and nights do not belong to them. The human joy or sorrow they know not. Yet man is the 'world' in which they live.

Earthman, the same way we live on Earth, but can we say that we truly know what this Earth is? For several years gone by, the Earth Planet has harbored the children of men solely in the spirit of 'Universal Love', as approved for her by the One Eternal Spirit. The reality and some hidden intricacies of this are unknown in the general human terms.

However, in the ubiquitous empty Space, the Earth Planet stood in LOVE to provide shelter for mankind to dwell; that man may advance in the great school of life; that man may perfect certain aspects of life that can only be achieved in the physical system of reality; that man may thereafter rise into higher subtle realms of Light on Earth beyond human sight, or into higher subtle realms of Light in other Planets, also beyond human sight.

## HIDDEN THINGS OF EARTH

There are so many hidden things about the Earth Planet that man does not know at the moment. But in the journey of endless life he will know them, for he will surely know all that he should know. At the moment man

Deeper Realities...

does not know in general terms that, apart from the physical realm of Earth on which he lives, the Earth has other realms harboring humanoid. There are realms of existence on Earth vibrating on diverse higher frequencies and diverse lower frequencies. Each frequency (realm of life beyond the physical) is separated from the other, not in terms of distance in miles, but by some degree of ethereal intensities.

Thus, the Earth has six higher realms, one physical realm (our present realm), one sub-physical realm (the immediate realm beyond the physical) and four lower realms. Beyond all these is the core of the Earth, the vital-force of the Logos of Earth. (This is the nature of all Planets). Each realm of Earth harbors humanoid whose consciousness dovetails with its frequency and vibrations.

For instance, all religions speak of hell. This we understand to be the lower realms. Jesus the Christ also visited these "lower parts of the Earth" (Eph 4:9 and Ps 63:9). In the higher realms of Earth are some paradises and heavens. Into one of the paradises Christ took the repented thief soon after His death on the cross (Luke 23:42,43). To show that the paradise He went is located in another realm on Earth, the Lord after His resurrection told Mary Magdalene: "Touch me not; for I am not yet ascended to My Father" (John 20:17). It means that after His death, He went to amongst others, a Paradise located on Earth, resurrected and thereafter ascended beyond all the realms of Earth. The Bible even reveals the existence of "man in heaven, in earth and under the earth" (Rev 5:3). Details of all these will surely take volumes of books. I will for now let it be, as it is not the subject matter here.

That the Earth is a conscious living Being harboring us, just as we are conscious living Beings harboring some minute creatures, is not a new thought in the world of man. Though many of us have not come to this realization. To that extent, many of us are ignorant of certain waves of interaction existing between the earthman and the Earth Planet.

In the Igbo cosmology, the Earth (Ani) is appreciated in every 'new-yam festival', after the traditional yam harvest. The Igbos do this in the thought that the Earth is aware (fully conscious) of their act of appreciation. The ancient Greeks recognized the Earth as a living conscious Being. They considered the primordial core of Earth's consciousness, the vital-force of Earth, as a Cosmic Entity of feminine gender, whom they called 'Gaia'. They even sang hymns to 'Gaia' as the 'mother of all' on Earth.

Deeper Realities...

## THE BASE OF MANIFESTATION

In one of the 'Vaishnava sastras' (scriptures) known as Srimad Bhagavatam, a part of the ancient Indian Vedas, (the first written holy books among men), we are told that the Earth is a full conscious living Being. In it we read of how the core of the Earth (the 'God of the Earth') went to the 'presiding God' of the Universe (our Galactic System) over some problems on Earth. The problem was the activities of some earthmen in conjunction with the most vicious Astral entities who were unsettling the physical and other realms of Earth with highly concentrated powers of darkness.

The 'God of the Earth' joined by some 'Interplanetary Gods' were led by the 'presiding Galactic Lord' (Brahma) to seek solution from The Supreme Absolute Spirit, The Father, The Supreme of all Supremes, The LORD GOD of 'Gods', The Infinite Unmanifest, who is the base of all the manifestations of 'Gods', Universes, Planets, All Power, Every Creature and all the realms of Cosmic manifestations.

To this end, they went to the boundary between the entire creation and the Infinite Unmanifest HEAVEN of GOD. This boundary is understood from the Vedic metaphysics to be the region of endless universal waters, the point from which creation (manifestation) began, the point which the Vedas hold that GOD in HIS Version as 'Maha Vishnu' (Eternal Spirit) covers the endless waters.

This understanding is also found in the Bible, where we are told that from the point of the beginning of creation, "the Spirit of God moved upon the face of the waters" (Gen 1: 1,2). However, the solution that the 'Gods' got from the GOD of 'Gods' resulted in the vanquishing of specific high forces of darkness then on Earth.

## EVIDENCE OF EARTH'S CONSCIOUSNESS

Also, the Bible has it that in the innermost core of the Earth resides the 'God of the Earth'. It says;

> "these are the two olive trees and the two candlesticks standing before the God of the earth" (Rev. 11:4).

When we hear of the "God of the Earth", it follows that there is the "God of Venus", the "God of Mars", the "God of Jupiter", amongst others. These are representations of the core identity of each Planet, the base from which consciousness springs and spreads upon the body of a Planet. Just as the Soul is the core base from which consciousness springs and spreads on the human body.

Mundane scientists are not completely left out of the understanding that the Earth is a conscious living Being.

### Deeper Realities...

As far back as 1785, a scientist considered as the leading light and father of geology, James Hutton, declared that "I consider the Earth to be a super organism". This was during a lecture he delivered in Edinburgh. In August 1986, a team of scientists (geologists) led by Dr. Azzakov (an atheist) made a discovery beneath the Earth, which got them and everyone bamboozled.

They were drilling the deepest hole on Earth of about 14.4 km beneath in Siberia; suddenly at a point the bore drill began to spin widely, revealing that the center of Earth is hollow, for which the geologists were dumbfounded. Furthermore, the innermost recess of that hole indicated temperatures (extreme heat) calculated to be two thousand degrees Fahrenheit (almost one thousand degrees Celsius).

Above all, they discovered something which made them to abandon the entire project in fear. They attempted to listen to the movements at different layers of the crust by highly sensitive microphones that were lowered to the drill gorge. What they heard changed the logically thinking scientists to shaking ruins. They heard several human voices screaming. Dr Azzakov, the Director of the project, said: "We could hardly believe our ears. Although one voice was distinguishably separate, we could hear the screaming of thousands, or may be millions, of souls in the background".

## EARTH CAN HELP YOU

No matter who or what such 'souls' are, by this and other related discoveries, it is evident even from the point of mundane scientific research that there is consciousness on Earth beyond the physical aspect generally known. The Earth certainly has secrets. In the Koran (2:34) God (the Almighty Allah) said:

> "Did I not say to you, I know the secrets of the heavens and of the earth, and I know what you reveal and what you conceal?".

There are secrets about the Earth and the higher 'Planetary Systems' (heavens) which are given to us to know. Such knowledge will add to our victory in the battles of life. Some secrets, however, are known only to GOD.

The Bible book of Revelation (12:15, 16) says:

> "And the serpent cast out of his mouth water as a flood after the woman, that he might cause her to be carried away of the flood. And the Earth helped the woman, and the Earth opened her mouth, and swallowed up the flood which the dragon cast out of his mouth".

### Deeper Realities...

The consummate understanding of the deep spiritual meaning here, which reveals the Earth Planet as a full conscious living Being, is very clear.

The Bible here shows the Earth as a high personality in the system of the Forces of Light. She rose against the whims of a high power of darkness in full consciousness of the 'flood' which the 'dragon' set upon her body against a messenger of Light. Also, the Earth is here revealed as having the conscious ability to 'help' with a 'mouth' that she can 'open' consciously to do a thing. All these can only be done by one who is fully conscious of himself and not by a mere mass of inert gross matter.

One day man will understand how the conglomeration of the multitudinous streams of consciousness culminate into one central flow of consciousness to form the body of a Planet. But for now and from here on, I will take up the matter of the subtle interaction between humans, certain subtle forces and the Earth.

CHAPTER 4

## SUMMARY OF THE MYSTERY OF LIFE

Life on Earth and within the entire Cosmic manifestations, is a compendium of mysteries. It is often said that life is a mystery. Yes, and in it we are all students. Every man is a student of the great life. In my terms, I am a student of life who 'knows nothing' yet. For in the light of what I should know, what everyone should know, not only for the moment but within the framework of eternal existence, that which I know now is 'nothing'. Nevertheless, remember that every 'nothing' is 'something' as 'nothing' does not exist.

### REST IN PEACE ?

Existence on Earth is not all about being born, then live for a while and thereafter go and rest in peace eternally after death. The phrase, 'rest in peace' for those that 'go to heaven' after death, to be with the Lord, is a relative and not an absolute reality. If you are not with the 'LORD' in the course of your physical existence, there is the

### Deeper Realities...

assumption (in strict religious terms) that you will not be with HIM after death. But if we agree that the LORD is Omnipresent, then this assumption becomes feeble in approximation to the reality of HIS being everywhere at all times.

Again, if you are 'with the LORD' on Earth, hoping to also 'meet HIM' after death, it becomes evident that this is a correlation of location and experience within certain intensity of manifestation. This is so because how you are with HIM now should serve as a clue of how you will be with Him in the hereafter, with the point of difference occurring only on the state of vibration between the physical and the subtle realms. Thus, to 'rest in peace' therefore, should not connote in our speculation the mundane aggrandizement associated with 'rest' in the physical system of reality.

It is just like a student in the university. He toils day and night, often without rest, in his academic career. After graduating successfully, those he left behind who are yet to graduate, can say that he has finished from school to 'rest'. In a relative sense this is true, because he will at least 'rest' from school work.

However, in the absolute sense of 'rest' this is not the case. He will then begin to face the realities and challenges of life after graduation. These in some cases include getting a good job (if possible), renting a house of his own, bearing the burden of dependent relations, getting married and rearing children, amongst others. This illustration is all about man's life from the physical realm of Earth to the life beyond.

## REALITY OF MENTAL ACT

If one comes to me, saying; "Harbinger of the Last Covenant, what do you hold as the summary of the mystery of man's life on Earth"? My response may not be difficult to comprehend. To this end, I would certainly say: "Every mental act, any mental act, no matter how minute, is a reality; and in this reality man is responsible". This is a fundamental truth of life that some of us have not realized.

Your mental act holds all the fabrics of your existence in this world and the world beyond. In it the pattern of your soul destiny is found. Your mental act is the summary of your emotion and desire, including your belief system, stimulated by imagination; all which are embedded in what we know as 'thought'. Your words and actions are rooted in your thought. Thus, thought is the only activity that is real in the world of man and in all the realms of life. Your thought which is the core of your mental act is the basis of your physical experience; after death it will remain the instrument that determines your movement and every experience, within the classification of emotion and desire.

Therefore when I say that "every mental act is a reality", it is far deeper than the mundane aspects assigned to it by humans. When this is understood, you will realize that it compasses everything about man and the Universe, everything about the microcosm vis-à-vis the macrocosm. Your mental act forms the basis of your existence as well as the degree of your experience, in what may or may not happen to you. This is a 'Universal Law' which even Christ Himself pointed out when He said to man; " As thou has believed, so be it done unto thee" (Math 8:13)

## DICHOTOMY OF INNER PATTERN

As soon as a mental act is initiated, its reality becomes evident within diverse spheres of existence. A highly intensified one, which stems as 'prayer' to God and fulfills certain Divine rules in Love, goes to God while at the same time being actualized in multidimensional systems. Some higher Beings based on their nature, (whether of Light or Darkness) feed from one of the two poles of human mental act. Some mental acts come into physical manifestation, some come into manifestation in the so-called realms beyond. Above all, each and every mental act is stored in the core of the great Being we know as 'Earth'. The foregoing is made clear in this treatise.

## Summary of The Mystery of Life

Now, your mental act, your inner pattern of life, constitutes the sum-total of your thoughts, words and actions. Your mental act interacts and interchanges, within certain range of intensity, with specific Universal Super-conscious waves embedded in our Solar System, under the 'Immutable Laws'. Within the system of man's mental act, which holds the fabrics of what you were, what you are, and what you will become, (by some degree of the application of your volition), the mental matrix enters the propensity of dichotomy.

## COSMIC EXPRESSION OF ONENESS

Consequently, the reality of mental act is split into two different poles of expression; one fixed in Love and the other in hate. From each of these poles, whatever you think, say, or do on Earth impinges upon this Planet in one way or the other. This is one of the lessons of the great life which man must learn. All things in existence, man, ant, tree, Earth, Solar System, Galaxy, etc., are interrelated via the Universal Super-conscious waves, which never ceases to express the Oneness of all things in GOD.

Through the expression of Universal Love, we humans can co-operate with God and HIS Forces of Light to manifest on the physical realm of Earth the

Deeper Realities...

Paradise that we seek. In our human family units, communities and nations, this can be done. It may be difficult but not impossible. Already the POWER that leads and guides this is manifested on Earth. If you do the opposite, you have yourself to blame.

Every mental act (thought, word, or action) of love demonstrated on Earth from the positive pole, shines forth as Light. This Light is recognized as such and for what it is, not with man's limited gross material visual perception, but with the subtle mechanisms of perception beyond the mundane.


## LIGHT IN MOTION

We must understand that there are diversities in the systems of perception including sight. For instance, the way man knows a tree via his visual perception, is certainly not the same way it is seen and known to a bird, a lizard, or an ant. The great Cosmic Beings, some of which men call 'Angels', or 'Gods', may choose to see a tree in human terms or in terms unknown to man. To such higher Beings, therefore, any expression of Love by man on Earth is seen as 'Light', in motion.

This Light is magnetized by such mighty Beings of a homogeneous nature, the Forces of Light, in certain realms of the Stellar System. They 'feed' from the Planetary animation inherent in this Light arising from the core of our Planet and, by the 'Law of Reciprocal Action', bestow the benediction that has always served the well being of humanity in diverse ways. Such things as healing, peace, progress amongst others, do flow into the world of man in this manner.

## RADIATION OF POWER

In the first place, each mental act that gives birth to either Love or hate is magnetized and stored in the innermost chambers of our Planet, the core of Earth. The Light arising from the mental act of Love, stored in the core of our Earth, works for the greater radiation of Divine Power on the Earth's life-force, for the well being of all forms of life on Earth including man.

Thus, every thought, word, or action of Love on Earth is the expression of Light of the One Eternal Spirit, guiding us back to the Source of all manifestations. This is the only valid expression in eternity, the only true religion in all the realms of life: to Love GOD in full consciousness and manifest it by showing Love to all humans, to all creatures.

Deeper Realities...

In The Everlasting Gospel (vol., 33 : 63) we read:

"Who is the man who says that God does not require anything from man? Can you see what God wants? God wants us to love Him with all our hearts, and with all our souls, and with all our minds."

In chapter 48 verse 7 we read,

"I preach nothing other than love. This is because I know that love constitutes the Kingdom of God. You have a task ahead of you to reach this love".

## EXTRATERRESTRIAL INTERVENTION

Furthermore, there is a specific special squad of the high Forces of Light that comes around the Earth Planet four times every year for the past 56 years. The cycle of this phase of their operation will last till the next one thousand years of human calculation. This squad is led by a high 'Solar Lord', or what we call 'Archangel', who resides in a certain realm of Light, a heavenly realm, located in the subtle aspects of the Mars Planet. They come into orbit about 1,550 miles above the surface of our Planet, with a mighty Extraterrestrial Satellite (a highly advanced Angelic Spaceship) that is far beyond the comprehension of even the most advanced scientist on Earth.

It is possible for the earthman to see this Satellite as a 'shining star' when it comes. But often it is surrounded by a shield of invisibility, brought about through the regular succession (rotation) of electromagnetic energy units (photons) in a three hundred and sixty degree arc upon the periphery of the magnetic protective shield, in which it can't be seen physically by humans. The specific mission of this squad of the Forces of Light, which operates strictly under the Cosmic Laws of the One Eternal ALMIGHTY GOD, is to magnetize and magnify each thought, word, or action of Love by any human on Earth, any time they come into orbit, via their Extraterrestrial Satellite.

## SALVAGING HUMANITY

Their magnification of each human act of Love, is by a factor of three thousand times. This is very serious! No act of Love by any human anywhere on Earth evades this whenever they come. It means that one act of Love when they come immediately becomes three thousand acts of Love. They do this to increase the potencies of Love and the radiation of same on the Planet Earth and on the earthmen. This is to strengthen Love over the highly accumulated radiation of the darkness of hate, which under certain 'Law' gives the forces of darkness the base to fight to ruin the world of man. In their terms, this is a specific contribution for the sanitization of the human race in Light.

This dimension of Inter-planetary intervention on Earth by the Forces of Light is brought about by the Personification cum Manifestation of the Highest Divine Spirit in Two Supreme Versions of HIMSELF now in all the Cosmic realms including the Earth. The high Beings involved here are not concerned with the religious mode of whoever is performing the selfless act of Love. To them what matters is the act of Divine Love, whether by a Christian, Muslim, Buddhist, Hindu, or whatever. In their terms, whenever or wherever an action of Love is located, in each period of their operation on Earth, it comes under the aforementioned magnification. Their major concern is the ascent of the entire human race in Light in the course of time.

The foregoing are some of the things that occur, though unknown to some earthmen, when man via his words, thoughts and actions sets Love in motion. Nevertheless, the opposite of all these comes into play within the paraphernalia of Planetary and human existence, when man via his mental act sets in motion the vicious cycle of hate. I will come to the details of this later.

## KNOW MORE

I have used 'Solar Lord' in reference to the entity in-charge of the Extraterrestrial Satellite that comes from the Mars Planet, though most religious scholars are not familiar with this title. But at least the Bible did say that there are 'Lords in heaven' and that 'to us'(that is in the art of worship) there is only 'One God' (THE FATHER) and HIS manifestation as Christ the 'Lord of Lords" (See 1Cor 8: 5,6, Rev 19: 16).

We also read of the 'King of Kings and Lord of Lords in the Bible. Now the 'Kings' and 'Lords' in this connection are not necessarily all on the system of the forces of darkness, never, though some are. There are several trillions of Planetary, Inter-Planetary, Solar, Galactic, Inter-Galactic (etc) 'Lords' and 'Kings' within the entire Cosmic spheres, including the realms of Earth, who are on the system of the Forces of Light. These are all led by the One Eternal King of Kings and Lord of Lords. Each of these reflects in certain ways and degrees the 'Prime Identity' of GOD. On a certain reality, they constitute the high 'gods' which humans should not worship, but must not revile. "Thou shalt not revile the gods". (Exo. 22: 28)

## CHAPTER 5

## MOST VICIOUS ASTRAL MANIPULATION

Every single thought, word, or action of hate arising from the earthman in the world of man, sets darkness in motion, from which the homogeneous entities also 'feed'. These entities are the vicious Astral forces. Darkness in motion, as a result of man's vicious mental act, burdens the core of Planet Earth with the Planetary equivalent of what in human terms is called 'disease' or 'illness'. The illness affecting man is often caused by infection. The 'illness' of a Planet, but in this case our Earth, also comes by infection; and this infection is hate. Under certain accumulated intensity, hate becomes an 'infection' on the life-force of our Earth. This is not a mere theory, but a high spiritual and metaphysical fact based on the Cosmic Laws.

## BURDEN OF A PLANET

Intercepted by the forces of darkness, the most vicious mighty Astral entities, hate in conjunction with

## Most Vicious Astral Manipulation

other related factors, serve as the major tool in manifesting pain, conflict and sorrow in the world of man. The practice of all forms of wickedness, hatred in all ramifications, high vicious occult manipulations, terrorism and witchcraft, are serious burden to the life-force of the mighty Being we know as Earth, the world in which we live.

Furthermore, the five principal acts that serve as the major channels of vicious Astral attack against humans, namely: lust, anger, greed, vanity and attachment to mundane things, all constitute a burden to the Earth Planet. All these that burden the Earth, having been practiced for so long in the world of man, at a certain stage sprang the great 'illness' of Earth. This is completely unknown in the mundane aspects of the earthman whether in his religion, science, or philosophy.

Not only these, for in the course of time, even our time, the tremendous amount of radioactivity brought about by the earthmen via atomic experimentations, joined to all the factors of wickedness, becomes the greatest burden of our Earth. This burden transmogrified into a "deadly virus" that was about to cause the death of Planet Earth; for then Earth was 'sick' and about to die, but how certain interventions saved the entire Planet will be made clear later.

However, without such intervention arising from the frequency of the 'Holy Spirit' Earth Planet would have

**Deeper Realities...**

been destroyed completely; that is Earth would have died. The earthman may be bamboozled by how it is possible for humans to kill the Earth. Of course it is possible for humans to destroy our world. Though the human form of life is very small when compared with the size of our Planet, under some circumstances humans can knowingly or unknowing sever the life-force of this Planet, and thus cause the death of our Earth.

## LITTLE CREATURES, MAN AND EARTH

To this end, when you consider the body size of an earthman and the size of a mosquito, it will become evident that a mosquito is minute by comparison with a man. Nevertheless, when mosquito bites man, it can release parasites that cause malaria, which is capable of taking life out of a human body.

In other words, a little mosquito can cause sickness, or even death, of the physical body of a human being. What of virus? This is also a living thing too small to be seen by man without a microscope. But regardless of the fact of its minute size, virus causes infection in people that could lead to the death of man in the physical system of existence. Thus, the size of these minute creatures amongst others do not determine what they can do to man

In the same way the size of an earthman does not determine the extent to which he could harm or kill Earth. The thoughts, words or actions of hate and their propensities, when accumulated in high intensity, form 'infectious disease' in the core of Earth wherein we live. In the Bible book of Revelation (11:15-18) we found that when the seventh Angel sounded his trumpet, among the things that were set forth to occur on Earth include the Divine decision to "destroy them which destroy the Earth".

Those whose activities work towards the death of our Planet, do so from the framework of the mind embedded in the frequency and propensity of hate, their mental act. In the first place, there are strong vicious Astral forces that strive constantly to invade the mental pattern of man, the human mind. This can be done against a person as well as a group of persons, for a specific result. But the success of the powers of darkness in this regard must follow strictly the 'Cosmic Law of Homogeneity' or the Universal Law of Attraction. All things in the Universes of God work within the ambit of the Inexorable Cosmic Laws.

## THE INVADED MIND

Therefore, for the high Astral forces of darkness to be successful in the craft of invading the mind of a person or

**Deeper Realities...**

group of persons, for the manifestation of a specific project, the person or group of persons must by 'law' be in a position to attract such high frequency manipulation. In this connection, the position required from man by the forces of darkness is that his mental act remains bereft of the 'Consciousness of Universal Love', which in reality is the 'Light of God' in motion. When it is said 'receive Jesus Christ into your life' or 'worship Allah and believe His Prophet' or 'worship Jehovah the only True God' or 'be born again', amongst others, I say fine. But you must understand that no matter how high or deep you think you are involved in any of these, if you fail to manifest selfless actions of Love, which set God's Light in motion, the vicious Astral forces will have their way when your mind is invaded.

When a man's mental pattern is invaded by the forces of darkness via the astral wave, his mind will begin to fluctuate and act like a monkey intoxicated by strong wine. The monkey being drunk staggers from fruit tree to fruit tree till he stumbles into a nest of scorpions, which swarm on him. As they sting him all over, he in great anger went about lamenting in pain.

This is the state of an invaded mind. The drunken monkey is symbolic of an invaded mind that endlessly seeks after mundane things, the material things of life. The monkey staggering from one fruit tree to the other, shows how a mental pattern bereft of Love, the Light of Life, staggers and wanders in the realm of the mundane seeking pleasure, the 'fruit of evil', which by the manipulation of darkness appears to be the 'good fruit'. Drunk with selfish desire, the human mind runs blindly into sense pleasure, the nest of scorpions. When the scorpions of greed, lust, pride and jealousy sting, lamentation follows and the mind filled with darkness enters the motion of furious hatred. At this point the Astral forces of darkness 'feed fat' from the mental energy arising from the thought of such a darkened mind.

## SYMBOL OF HATE

Earthmen, you must know that the greatest evil which the Astral forces of darkness used you to manifest on Earth, via the invasion of your mass mental pattern, is the invention of the atomic bomb. This is the greatest symbol of hate and evidence of wickedness of the highest order. The thought pattern of man that served as the base of this, represents the vicious mental act of a high dimension.

It was a thought pattern that arose through the accumulation of hate in the 'psychic-bank' of humanity for several past generations. This was consequently hijacked by the alien mighty dark powers of the Astral realms. These are highly intelligent and powerful forces

Deeper Realities...

of darkness. Their sole aim is to mutate humans via atomic radiation and use the mutated and distorted consciousness as a strong base to battle the Forces of Light in our Solar System. The project to mutate humans in their terms will make them to have full dominion on Earth via the Immutable Laws, whilst the physical Earth is set apart as the world of man; and to them it does not matter even if the Earth dies in this regard.

To this end, the forces of darkness deeply concentrated in the art of high manipulation for the manifestation of atomic bomb on the physical system of reality. Their project went through, as they used highly advanced human scientific centers to bring this to pass. We must understand that it was man's collective mental pattern, within the framework of hate, that gave the forces of darkness the lawful base for this.

Furthermore, the forces of darkness also used such highly advanced mundane scientific centers, via some human channels as always the case, to bring forth other things. For instance, they developed disease spores for bacteriological warfare, as well as the invention of hydrogen bombs on the physical realm of Earth, amongst others. All instruments of human military warfare designed to kill the children of men, came to the physical Earth from the vicious realms of Astral forces.

However, as already stated, the manifestation of these and much more, were made possible by the vicious human mental act, through which man was also led by the dark forces to poke into the realm of atoms. Till now, some of us do not know that atomic bomb is far more dangerous to humanoid, even more dangerous to the life-force of Earth, than HIV is to humans on Earth.

## MUNDANE LIMITATION

Some years back, still within the frequency of the great Astral vicious plan to destroy our Earth, though unknown to some humans involved, there was a general conclusion that nuclear tests or accidents are harmless to humans. This dimension of thought coming from the masters of the vicious Astral forces in the realms beyond, was promoted on the physical realm of Earth by highly placed politicians and scientists. Such earthmen then did not understand the full danger and high negative effect on humans and the life of Planet Earth regarding the 'escape of radioactive iodine and strontium 90, as well as plutonium dust', in the nuclear device.

The statement once officially made by Harold Mac Millian on the 8th of November, 1957, is very important in this connection. He was then the British Prime Minister. He said among other things that "There is no evidence that this accident has done any significant harm to any person, animal, or property." The 'accident' he

Deeper Realities...

spoke of was a nuclear mishap at Sellafield on the 10th of October, 1957. What he said, however, which he made in the 'House of Commons', was not a mere personal opinion. It was strictly based on the findings of two special committees set up to investigate the nuclear accident. The first was a committee of foremost scientists set up by the 'Atomic Energy Authority', while the second was a committee of medical experts set up by the 'Medical Research Council'.

What he said at that time encapsulates the general thought of the earthmen on the mundane level, mainly promoted by leading politicians, high medical experts and prominent scientists. Following this, the testing of atomic device was considered 'safe' and 'normal' by the so-called 'world powers'. But in the course of time, these earthmen were bamboozled on what has become the adverse effect of atomic experimentations.

For instance, several Sellafield workers later became victims of cancer in the aftermath of the nuclear accident, for which the British Nuclear Fuels Ltd. paid huge compensation to their families. Hitherto, medical experts have also confirmed that leukemia in humans living at Seascale near Sellafield, especially those born since the accident, is far higher than in humans born in other places.

## FACING THE REALITY

Also in our time scientists have come to terms with the reality that underground nuclear test piles up high radioactive energy, which seeks weak spot on the Earth Planet's crusts for catastrophic expression. When it finds a weak spot, the radioactive energy goes into manifestation in the form of earthquake or volcanic eruption. Thus, scientists at New Brunswick University, Canada, have precisely established a link between underground nuclear tests and earthquakes/volcanic eruptions.

There has never been a time that nuclear test or accident occurs without being followed by either an earthquake or volcanic eruption after a while. When France tested a nuclear device in the Pacific in 1985, it was soon followed by earthquake in North America. Another nuclear test ten years later by France also resulted in earthquake and volcanic eruption thereafter. The great earthquake that killed thousands of people in Armenia occurred soon after the Soviet nuclear test. Even the Los Angeles earthquake of 1994 took place not long after a nuclear test conducted by the United States in Nevada, amongst others.

For now, mundane scientists already know that any explosion of atomic bomb must result in heat and radioactive waves. They have also noted the fact that

> Deeper Realities...

major meteorological changes have taken place via atomic releases. They are still researching to know the full reason for this. But mundane scientists on Earth are totally ignorant of the fact that atomic explosion also gives birth to high etheric distortion. This is a fact well known to the mighty Forces of Light in the Higher Planetary Systems.

These mighty Angelic Beings of the Almighty in the higher realms of Light, have revealed that etheric distortion does not only come via atomic explosion but also by intensified vicious human thoughts, which cause tension on the etheric atmosphere and manifest on the physical Earth such things as floods, famines, droughts, to mention but a few. I will still come up later with other deeper realities connected to atomic radiation completely unknown to our earthly scientists.

## CHAPTER 6

### DIMENSIONS OF EXISTENCE

Contemporary mundane science does not comprehend how human vicious thoughts accumulate in a subtle system and thereafter manifest physically in the form of natural disasters. When a boil, 'a localized, inflamed swelling of the skin', manifests on the human body, we understand that it is 'caused by a bacterium' somewhere inside the body. In the same way, the human thoughts embedded on hate and every form of the act of wickedness, constitute the 'bacteria inside the Earth' from which we could have the manifestation of such destructive catastrophies as fire blazing through forests, volcanoes, ocean mega-waves, earthquakes, hurricanes, tsunamis, etc.

This may be difficult for the present man to understand because it is based, not within the framework of mundane religion and science, but on High Metaphysics confirmed by the mighty Forces of Light in the heavens of Higher Planetary Systems. These mighty Ones are advanced Extraterrestrial Scientists whose science lies millions of years ahead that of humans. These highly advanced Extraterrestrial Scientists, are 'masters' of the science that has totally realized the existence of basic subtle matter in the entire Seven Dimensions - Height, Breadth, Length, Time, Motion, Mind and Will.

## CONTROLLER OF MANIFESTATIONS

The 'Seventh Dimension', which is 'Divine Will,' is the 'Controlling Force' of all manifestations, all realities, all existences. This is the Dimension that we call God, Jehovah, Vishnu, Holy Spirit, Allah, Father, Christ Consciousness, Chukwu, amongst others. Mundane science is bereft of any framework that could lead to this Dimension, but in general terms religion has, at least by recognizing Its existence in theory, so also is metaphysics. However, what we call 'High Metaphysics' has gone far beyond both religion and regular metaphysics on this matter. It has through contact with Extraterrestrial Intelligences in the Higher Planetary Systems, who somehow or the other connect to the Cosmic realms of the Seventh Dimension, revealed great facts and deeper realities of existence to a greater extent.

When we study the sacred writings of world religions, we will discover that they were given by God to guide man via great Extraterrestrial Beings who connect into the essence of the Seventh Dimension.

The Koran (2:99) has it that:

> "Whoever is an enemy to Allah, and His Angels, and His Messengers, and Gabriel and Michael, then surely, Allah is an enemy to such disbelievers".

My point here is the mention made of 'Angels, Gabriel and Michael.' While the Bible also speaks of 'Angels, Cherubims and Seraphims, Watchers, Archangels, Living Creatures, Elders, amongst others, the Veda speaks of Manus, Kumaras, Sanaka, Sanatana ,Sanat, Mahajanas, amongst others. Even in our time, direct contact is made by some earthmen with the 'Cosmic Brotherhood of Space Masters' and much more.

The aforementioned Beings are not of the terrestrial (earthly) life, but of the Extraterrestrial and Celestial existence. Some of them operate fully from the Sixth Dimension, while some have established levels of connection into the framework of the Seventh Dimension. Some humans now on the physical Earth are the manifestations of these. Thus, while such ones live in the world of man, their consciousness transcends the lower dimensions. Not only these manifestations, for there are also some humans still on Earth who have risen to higher consciousness via what we may call 'divine promotion'.

*Deeper Realities...*

This Seventh Dimension is highly variegated in levels as well as limitless in eternity. If a Consummate Personality manifests on Earth from the summit of this consciousness, then it literally means that 'God has come down to dwell with men', or that the 'kingdom' of God has come. The case of Our Lord Jesus the Christ is very important in this connection. Even in our time, there is a 'high metaphysical' evidence of this on a larger scale, which will in the course of time completely overwhelm the entire world of man. This is in fulfillment of the ancient prophecies in God's terms, not in human cum religious terms.

## OTHER DIMENSIONS

Mind is the Sixth Dimension, which again is highly variegated, from the turbulent lower levels to the subtler super conscious levels. The mundane science of Earth is far below the Sixth Dimension, which is the basis of the science of higher civilization in the heavenly Planetary Systems. The Seventh Dimension, the base of all manifestations that is also above and beyond all manifestations, is the 'One Eternal Spirit'. This Dimension dictates all paraphernalia of manifestation and acclimatization, including the conditions of these, upon the Sixth Dimension, Mind. Consequently, the Sixth Dimension exerts its potential into expression as regulated 'Motion' (Fifth Dimension) within the 'Time' (Fourth Dimension) sequence of (Three Dimensions) Length, Breadth and Height.

## INFLUENCE OF MIND OVER MATTER

Thus, the Sixth Dimension (Mind) has dominion over the elements of matter. This is what mundane science does not know. Earthly scientists are far below the understanding of how thought from the mind dominates and exerts influence over matter and the elements. At the moment they cannot understand 'why' and 'how', as well as the 'link' between man's vicious thoughts and the manifestation of 'natural disasters'. Even what they know as 'atomic bomb', they do not know what it is in all ramifications, whilst this in itself is the by-product of vicious mental act completely bereft of the recognition and consciousness of Love.

The atomic bomb working hand in hand with the accumulation of the mental pattern of hate, is all that the wicked Astral forces want for the annihilation of all life forms on Earth. Already it is well known that the amount of atomic bombs on Earth can destroy all life forms on this Planet even six times. When we consider this in the light of constant flow of wrong thoughts from humans into the ether of our entire eco-

Deeper Realities...

system, then it becomes evident that, in the absence of any higher 'Divine intervention', the earthman has already set the stage for 'the end of the world'. To this end, 'Great Ones' in the higher Planetary realms of Light, the Forces of Light, have seriously warned against the 'release of electro-magneto-motive power,' which is set in motion 'from the bombardment of unstable uranium 238 under high vacuum by slow neutrons'. This has tremendous consequences on the ether, the weather and the entire life-force of Planet Earth and beyond. The same goes with human thoughts embedded on hate in diverse forms of wickedness and terrorism.

## A STRANGE FACT

The scientists of our world are completely ignorant of a fact, fully known to mighty Beings in the higher realms of Light that the negative effect of atomic radiation goes beyond the physical aspects of life. Mundane science does not know that the effect of radioactivity goes into all the subtle realms of Earth, such as the heavens, the paradises, amongst others. Neither does mundane science understand the great danger of the subtle aspects of atomic radiation both for terrestrial and Extraterrestrial existence. This knowledge coming via 'High Metaphysics' goes beyond the present understanding of the earthman in all his mundane sciences, philosophies metaphysics and religions.

When a civilization in any Planetary system of the Cosmos reaches the point of poking into the atom, the point where man is now, it will stand upon the threshold of winnowing. Some from this point used the atomic energy for positive ends. Consequently, such civilizations transcended, enriched and embellished their worlds. At the moment, such civilizations have become completely spiritualized and transmuted into what the earthman will call the 'Kingdom of God'. The Earth can also become the 'Kingdom of God' when we all comprehend the practical imperatives of living in Love.

However, some civilizations within some Planets also reached the present point of Earth, poked into the atoms, invented atomic bomb, used same severally in a 'world war', destroyed every life in their Planets including the Planets themselves. Some even entirely exploded their Planets. This has happened also in our Solar system. The Extraterrestrial Intelligences in the heavenly Planets have for long confirmed that what our Astronauts on Earth know as 'Asteroid Belt', between Jupiter and Mars Planets, was once a Planet that harbored humanoid with physical life, as we have on the physical realm of Planet Earth. This Planet was the last

Deeper Realities...

destroyed in our Solar System by her inhabitants via a nuclear holocaust in conjunction with high practice of hate and the most vicious forces of terror.

Even in our time, some scientists who previously held that the Asteroid Belt is the left over material from the composition of the Solar System, have retraced their theories. Some of these, including Professor Ovenden (former British Astronomer), after over twenty years of research, now consider this as 'debris of exploded Planet'. There are high and diverse efforts to avoid a repeat of this, though unknown to several humans.

## VICIOUS THOUGHT PROJECTIONS

In this treatise I have tried as much as I can to make it clear that the Earth, as a conscious living Being, was sick at a time and about to die, not quite long ago in our terms. This was brought about via wrong or vicious thought projections of the children of men embedded on hate. We all know that everything begins in thought. Words and actions are expressions of thought. When a high vicious action is performed in any part of the physical Earth, the thought intensity of that moment-point settles as a 'virus' in one of the Earth's psychic centers, rising from there to the core of Planet Earth. So also is the action of Love, which settles to radiate Light.

## Dimensions of Existence

Now, a psychic centre is the specific location from which the Life-Energy flows into the body of a Planet, into the body of a man, or into the body of any creature. There are diverse of such located on the physical realm of Earth in different parts of our world. There are specific vicious thoughts, words and actions that are highly detrimental to them, which however serve as 'food' to other kind of Beings, the Astral forces of darkness.

Consider the acts of rape, abortion, murder, arson, armed robbery, kidnapping, fighting, greed, war, revenge, quarrel, intimidation, exasperation, terrorism and a host of others. At the point any of these is enacted, a high level of vicious thought projection is set in motion. Even when the individuals involved seem to have forgotten, or feel it is all over, such thought continues to work in other dimensions, completely unknown to the ones who initiated them.

All these having been practiced for ages in the world of man, constituted the great 'illness' that almost took the life of our Planetary Mother, the Earth. And when joined by atomic experimentations, arising from the vicious mental pattern of humans, the stake was raised for the total annihilation of Earth. It is estimated that about fifty million people died in the 2nd World War. During this war, atomic bombs were used on the 6th and 9th of August, 1945, respectively, though on a small scale. If used on a large scale as we have now, with the more dangerous hydrogen bombs that came up from 1952, then humanity should forget our world.

## UNDERSTANDING THE RELATIVITY OF NATURE

To overcome this greatest danger, there are diverse high frequency Divine activities going on in the higher realms of Light. Every man also has a part to play in this connection. Your slightest mental act is very important in this regard, your slightest thought, word, or action of selfless Love. Our efforts must be geared towards this with utmost seriousness and commitment, bearing in mind that there are high forces of darkness also highly committed towards thwarting every effort of Love.

From the point of the Astral forces of darkness, they work very hard in the consciousness that they are fighting for their well-being and survival. Part of their 'food' includes the 'ripe fruits' of human vicious thoughts; and they fight hard so as not to be denied this. You see, a lion fights hard, or runs fast, to catch his 'food'. Now, this food could be an antelope. If well armed men delivers the antelope from the lion, it follows that, from the view point of the antelope, the lion is 'a power of darkness', while the armed men are the 'forces of light'.

The same situation applies to a tree. There are some humans who cut down trees to eke a living. There are also some humans empowered by the government to stop the cutting down of trees. From the point of view of a tree, those cutting down trees are 'forces of darkness' (even when their aim is strictly for human survival) and those preventing the cutting down of trees are the 'forces of light'. That is how it goes in nature.

We must understand that the fruits we eat are 'given birth' by living creatures full of life. They may be lesser creatures but of course they are still living entities, just as man is also a living entity on a higher scale of existence than them. In some sense we found in nature the process of higher creatures feeding on the lesser ones. Though man in one sense is a higher creature, but in another sense he is a lower creature from whose 'fruits' certain higher creatures than him feed on.

Deeper Realities...

## CHAPTER 7

### THE GAME THAT SAVED

We must know that all human thoughts within the framework of hate and wickedness, most especially the mental act of bearing grudge, on a certain degree of intensity, become to the mephistophelian Astral forces what in human terms are 'ripe fruits.' They practically feed on these for greater psychic strength, in the same way that the earthmen eat food for physical strength and earthly survival. They work hard day and night to ensure that the process of having the flow of their 'food' from humans is not cut off by the Forces opposed to them.

This they do via the constant invasion of the human mind. In this, however, lies a great battle: The battle of survival! The forces of darkness versus the forces opposed to them - the Forces of Light. In our Solar System, the Earth Planet, for some Interplanetary reasons, is both the 'primary school' of humanoid as well as the current 'war zone' of the Forces of Light and the forces of darkness.

## SURVIVING ON HUMANS

The 'mighty ones', mighty in the arts and crafts of darkness, have the great occult powers and high psychic capacity to hijack the slightest human vicious thought. Thereafter they can split up such a thought form and feed fat from the mental energy contained in it, which in their terms is but a 'ripe fruit', the way we humans feed from the fruits of some creatures. Apart from this, they can and often do 'plant' such mental energy as a major tool to further invade the mental pattern of men, for the growth of more food in the 'psychic bank' of humans. Remember, we humans also plant crops for the growth of more food.

As the high forces of darkness use the vicious thoughts, words and actions of humans for higher manipulation, as edibles and basis of astral attack against humanity, so also the Forces of Light use the human thoughts, words and actions of selfless Love for another purpose. Their main purpose to this end is solely benevolence for man, strictly for the ascent of the children of men into the realms of Light. They work day and night and strenuously as well to reduce, or possibly empty, the food in the 'psychic bank,' which the earthman do put in stock for the forces of darkness, via their wickedness. The Forces of Light, the Mighty Ones of the Almighty, fully know that the more food the forces of darkness have in the 'psychic bank' of humanity, the more 'legal right' they have to endanger the human race and Planet Earth.

Deeper Realities...

## SIGNET OF MYSTERY

In connection with this, therefore, high multidimensional efforts in different realms of Light are in motion, in different ways, to checkmate the forays of the Astral vicious forces. These vicious forces themselves are not relenting. They are working very hard every moment to bring about the necessary condition, via the earthmen themselves, to fully possess the world of man. To them this is the number one duty of all duties. Humans are generally oblivious of all these. Rather they continue to empower darkness against themselves and often vilify the bona-fide high channels of Light who are on Earth to lift up the human race.

Every effort, whether by the human high channels of Light on the physical realm of Earth, or by the Forces of Light in the realms beyond the physical, to salvage the Earth now and the entire human race, is directly coordinated by the Supreme Absolute Spirit, the Spirit of all Spirits, the Holy Spirit, the Highest Divine in Two specific Versions of HIS manifestation in the world of man. These Supreme Versions are also Personified and manifested within some 'duty-post' in all the realms of every Planet, Sun, Moon and Star of the entire Cosmic manifestations. In our time, this is the consummation of the mystery of mysteries.

The Interplanetary effort of the special squad of the

Forces of Light, coming from a heavenly realm in the Mars Planet, as I earlier stated in this treatise, is very important in this regard. Within a specific programmed period of four times a year, each of these lasting not less than four weeks, they come to assist man in their own advanced way. With a highly advanced Extraterrestrial Satellite, as empowered by GOD at their level, they set out in each period to magnify any single thought, word, or action of Love, by any human on Earth, into a factor of three thousand times. This goes a long way to decrease the aforementioned 'psychic bank' from which the forces of darkness in the Astral realms feed. As this and other efforts continue, via the radiation of the Holy Spirit, the entire physical realm of Earth will surely and soon become the realm of Light.

## ASTRAL RIGHT

When Our Great Lord Jesus the Christ manifested on Earth, the food stored for the forces of darkness in the collective psychic bank of humanity, via the wicked mental pattern of the human race, had become extremely large. This gave the forces, elements and agents of darkness the necessary 'astral legal right' to have full dominion in the world of man.

This brought about a colossal dimension of battle

Deeper Realities...

between the Forces of Light and the forces of darkness. Christ led the Forces of Light with the sole aim of taking back the human realm, for the 'Divine Government' of the One Eternal Spirit. In doing this, a high Divine game was worked out by the Lord Himself to defeat the host of darkness. To this end, the Lord undertook a 'strange cosmic mission,' in which He allowed Himself to be crucified (Shed His Blood) in the world of man by the forces of darkness, via their principal human agents.

## COSMIC CRIME

By working out the manipulation which led to the murder of 'The Son of Man' in their terms, but fulfilling the script and game plan of Supreme Light unknown to them, the entire forces of darkness committed what the 'Universal' Immutable Laws' recognize as 'Cosmic Crime'. To deride the Lord, revile Him, let alone killing Him, when he manifests on Earth, in a human form, is a great 'Cosmic Crime'. It is within the system of what the Vedic Metaphysics consider as 'mad elephant offence'.

In the Bhagavad-Gita

> (a holy book written in 3000 BC) God said: "Fools deride Me when I descend in the human form. They do not know My

transcendental (divine) nature as the Supreme Lord of all that be"

(B.Gita 9:11).

To deride the Lord in manifestation, blaspheme Him, or work against Him in one way or the other, is a serious crime that goes with great consequences on the offender in the course of the journey of life, even beyond the physical realm of Earth. To also maltreat, fight or harm a truly empowered 'Messenger of Light' in the world of man, is equally a serious crime that brings punishment. In the Bible, God said: "Touch not mine anointed, and do my prophets no harm" (Ps. 105:5).

A great Divine game is constantly worked out by the Supreme Spirit of Light in all ages to assist the Forces of Light defeat the forces of darkness, in the battle which the Lord coordinates for the sanity of existence. This game is often beyond the purview of the framework and doctrines of organized religions. The game is this: the Lord who in one sense prohibits the crime of deriding Him in manifestation, or His bona-fide messengers, also permits the forces of darkness in another sense to commit the same prohibited crime. Why? So that through such 'mad elephant offence' a ground of great punishment will be established by the Forces of Light against the forces of darkness.

## BASIS OF PSYCHIC ATTACK

The major punishment to this end against the host of darkness, however, is what in human terms we would call 'starvation'. That is what it is in the strict sense of the word. I have said that the vicious human mental pattern that gives birth to hate and all forms of wickedness, is the major food of the forces of darkness. I have also said that this food is contained in the collective psychic bank of the human race. Now, the more food the forces of darkness have in this regard, the more they feed fat, the more grip they have on humanity, the more astral right they have to dictate what happens in the world of man. This is strictly in accordance with the 'Immutable Laws' that govern all the realms of existence.

In other words, the more they are fed by humans, the more they have the 'astral legal right' to do what they want with humans. Cogitate the invention of atomic bomb, the coming up of such a human as Adolf Hitler and his type, the high profile network of well coordinated terrorism on Earth, vicious wars, genocide, ethnic cleansing, to mention but a few. All these came into manifestation on the physical realm of Earth via the manipulations of the malicious Astral entities, based on the degree of empowerment they had from mankind, (their food in the psychic bank of the children of men).

## MAN'S PRIME DUTY

Every thought, word, or action of selfless Divine Love among men 'strengthens' the Forces of Light to also have the 'divine right and legal base' to fight for humans and reduce the food in the bank of darkness. This should be our individual and collective duty now. Nevertheless, when the forces of darkness through their human channels commit the 'mad elephant offence' of maltreating the Lord or His Messenger, one of the major consequences that follow is the large banishment of what they feed on. Thus, from time to time, they are tricked by the Lord into committing this offence.

When man plants food crops in his farm, he certainly would not wish to host a mad elephant there, so that the fruits of his labour will not be wasted. If a mad elephant enters into a farm, the insane beast will destroy the food crops and possibly lay waste the entire farm. This is what happens when an agent of darkness fights the Lord or His Messenger, it becomes a 'mad elephant offence' that wastes the 'fruits of their labour'. This method has been used greatly by the Forces of Light to defeat the network of darkness in the field of battle.

Deeper Realities...

## THE GAME OF JESUS THE CHRIST

This was the same method, though on the highest dimension, which the Lord Himself, in His manifestation as Jesus, used to bamboozle darkness and thus saved the world. He deliberately allowed Himself to be killed by the agents of darkness. He tricked the forces of darkness to do this. He beats them hands down in their own tricks and manipulations. They 'killed the Lord', though the Lord is beyond birth and death, but He accepted it which in His terms fulfilled His higher serious game plan.

By the time the Lord died and resurrected from the dead, it became evident in all the Cosmic realms that He is the manifestation of God on Earth; and that the forces of darkness who killed Him have committed the worst 'Cosmic Crime', the crime of all crimes, the mad elephant offence of all mad elephant offences. And by the eternal Immutable Law of Reciprocal Action, they had to pay several fold for this everlasting crime. Their payment for this great crime was, among other things, the complete work of the mad elephant in their farm, the psychic bank of the vicious works of men from which they feed, and on which they possessed the right by law to take over the Earth.

This bank, which feeds the forces of darkness, that is every food in it, was completely erased, following the death of 'The Son of Man'. This also came by law, the Law of Reciprocal Action. But in this case the operation of this law was magnified into a factor of several folds, because the crime in which it took a full course was a Cosmic Crime, the most vicious crime of all crimes, the murder of the Godman.

## SAVIOUR OF THE WORLD

Consequently, the death of Lord Jesus the Christ, via the shedding of His Blood, became the Sacrifice of all sacrifices; the Supreme Sacrifice of the Almighty God, made by God Himself to salvage the world of man. As His death erased all that the forces of darkness had then as their basis of subjugating the entire world of man, the human race consequently began existing afresh upon a new slate on the physical realm of Earth. By this, Jesus the Christ became the Universal Saviour of the entire human race. He did not come to form a new religion, or to establish a church, but to save the world, which He did.

We must understand that the killing of Jesus was purely deliberate on the part of the Lord. He knew what would be the final outcome of this, as well as what mankind as a whole would ultimately benefit from it. In some instance He even deliberately conducted Himself in ways that stirred the agents of darkness to plot His murder. This was a High Divine Game of the Forces of Light, which He led and leads hitherto, played on the intelligence of the forces of darkness. A game plan which dragged them to lay hands on the Lord, to bring about the multidimensional retribution for this.

Therefore Our Lord Jesus the Christ truly saved the world from what is often termed 'the bondage of sin'. He is therefore the Saviour of every man or woman in all the realms of Earth. Whatever saves man also saves the lesser creatures under man. Thus, Christ is also the Saviour of every ant, bird, tree, fish, rock, beast, etc. He also proclaimed Love as the basis of maintaining His great work of salvation in the world of man.

# CHAPTER 8

## ETERNAL VALIDITY OF ONENESS

The very fact that Love is and will continue to be the final basis of Planetary and Interplanetary existence, was well known to the wise Ones in all ages. This fact is evident and remains unquestionable in all the ancient sacred writings and in all the sacred writings of our time. The problem has always been with the earthman on the practical application of this, for the sanity of his life and the world in which he lives.

## HIGHER PATH OF LIFE

Selfless thoughts, words, actions, including sincere prayers of Love, are what we (the human race) need now than ever. This is the framework through which the One Eternal Spirit will continue to radiate HIS limitless Power towards the transformation of the physical realm of Earth as the Paradise we seek. Selfless Love entails that, if need be, you hurt yourself instead of hurting your fellow man.

*Deeper Realities...*

The Word of God as found in the Divine Iliad has it that:

> *"Serve first thy brother. Hurt first thyself rather than thy neighbour. Gain naught from him unbalanced by thy giving. Protect thou the weak with thy strength, for if thou use thy strength against him, his weakness will prevail against thee, and thy strength will avail thee naught".*

The practice of Divine Love is the foremost evidence that the practitioner is on the higher strata of God-realization. The practice of Love, not religious affiliation or membership, is the prime evidence of God's Light in the word of man. This is so because, as Saint John said in the Bible, *'God is love'*. He wrote:

> *"Beloved, let us love one another: for love is of God; and everyone that loveth is born of God, and knoweth God. He that loveth not knoweth not God; for God is love"*
> (1st John 4: 7, 8).

## ESSENCE OF RELIGION

This is a high truth: 'Everyone that loveth', regardless of his religious membership, church denomination, or otherwise, 'is born of God'. The conclusion of this, according to Saint John, is that such are the people that 'knoweth God'. This is so because Love in thoughts, words, or actions, is the bonafide principle that sets in motion, for manifestation on the physical realm of Earth, the Light of the Almighty God.

Even the Great Lord Jesus the Christ Himself made it clear that the demonstration of Love, and nothing else, is the sole evidence of recognition of His true follower, one that follows the path of Christ. He said specifically on this subject matter:

> "By this shall all men know that ye are my disciples, if ye have love for one another"
> (John 13:35).

In other words, the Spirit of Christ, the Consciousness of Christ, is the legitimate 'Power Source' that generates LOVE in the entire Cosmic manifestations. All men that put Divine Love into practice in all ages, even to this moment, are touched and impelled to do so by the Spirit of Christ, whether they know it or not on the conscious level of existence. These therefore are the true channels, instruments, or 'followers' of the 'Everlasting Christ'.

Deeper Realities...

## TRANSCENDING THE MUNDANE RELIGION

This is a Universal Truth which some churchmen would not accept. They may insist that any act of selfless Love not performed by one that seeks God, or knows God, in organized church terms, can never lead such to God's Paradise. Though I am also a clergy in Christian religious terms and fully understands the inner psychology of this mental act, the doctrines and basis of this belief-system, but I think the hour has come for us to comprehend the deepest realities of the Manifestation and Message of Our Lord Jesus.

The Koran says:

> "And they say, 'None shall ever enter Heaven unless he be a Jew or a Christian'. These are their vain desires. Say, 'produce your proof, if you are truthful.' Nay, whoever submits himself completely to Allah, and is the doer of good, shall have his reward with his Lord. No fear shall come upon such, neither shall they grieve"
> (Koran 2:112,113)

. The true understanding of 'submits himself completely to Allah' here, is very important in this connection. Also important is 'the doer of good'.

Complete submission to God is actualized by loving God with your entire consciousness. This was the first commandment of God given to the Great Prophet Moses. The evidence of this submission encapsulates 'the doer of good', that is the practice of selfless actions, words, or thoughts of Divine Love. Without the demonstration of Love, there can be no submission to God. When one is fixed on the principle of Love, he disturbs no one and is disturbed by nothing. As God said in the Bhagavad-Gita:

> "He for whom no one is put into difficulty and who is not disturbed by anyone, who is equipoised in happiness and distress, fear and anxiety, is very dear to Me" (B.Gita 12:15).

## THE JOURNEY TOWARDS GOD

The earthman must eschew all disagreements based on which religion is the best, which church denomination is the best and all that. He must eschew all impediments on peaceful co-existence based on Love. This is my message. He must understand that all things, all men, are on the journey towards God, towards God's realization. In this journey, some are on the path that is very rough, very difficult to tread on, but very short. This

*Deeper Realities...*

is the way on which Divine Love leads. While some are on the path of the easy life, a very long path. Yet all men are on the way to God, just as God said in the Divine Iliad: "All men will come to Me in due time, but theirs is the agony of awaiting".

The 'agony of awaiting' is the 'fire of hell' which lies on the very long path. Even from this 'eternal fire', man must reach God, realize God as he should, in the course of time. This is another Universal Truth that some religions would never want to be part of. The Hindu, Buddhist, Christian and Islamic religions, amongst others, all speak of 'hell fire' for the sinful earthman in the after-life. This is also a high metaphysical fact, but in a relative sense.

The realms of hell are to the soul of man and the Almighty God what prison custody is to the body of man and the government of a nation. The man in jail considers it as punishment, because his movement is restricted by law. In a sense it is so, but the aim of government in setting up the prison house is reformation not punishment. Though the prison house ever remains, a man can go there and come out after some time to become more useful to himself and the society at large.

The same thing happens to a man who goes to 'eternal' hell fire after death. We should understand that hell is 'eternal' in two aspects. First is the fact that it is always there. Second is within the intensity of predicament.

What we know as the 'lower realms of Earth' are actually hellish regions of different levels and kinds. Certain people go there after physical death for reformation, to work out problems associated with character, which in another sense is 'punishment'. What takes people to hell is mainly lack of Love, lack of Christ, lack of God's Light, lack of God.

## THE ONLY WAY TO GOD

Jesus the Christ said:

> 'I am the way, the truth, and the life: no man cometh unto the Father, but by me"
> (John 14:6).

It is said that 'God is Love'. Lord Jesus is the Consummate Manifestation of this God, this Love. Therefore He as Love is the only way to God. No one can reach God, no one can realize God, without Love. That is the purport of this saying of Christ. If you die on Earth without Love, you will surely go to hell. But this does not mean that one who lived for about twenty years, dies and goes to hell, will be there for millions, billions and trillions of years, or even more. There is nothing like that in reality. GOD does not work that way. HIS Universal Laws do not work that way.

Deeper Realities...

The word 'eternal' used in connection with the existence of hell as the realm of 'eternal punishment', is relative not absolute. The relative nature of this is within the range and intensity of 'pain' in the framework of the absence of time. The absence of time at any given moment in existence on man's experience, brings 'eternity' into focus within a 'moment point'. In a state of extreme pain, time becomes absent in our consciousness, whilst at that point we experience nothing but 'eternity'.

If you put your finger in the fire for five minutes, then you will experience 'eternity' during this period; you will be practically bereft of the consciousness of time and become fully conscious only of pain. In this consciousness of pain bereft of time, what then holds forth? Eternity! Within that five minutes you have 'experienced eternity'. Can you put your hand in the fire and under serious pain begin to count seconds and minutes? Never. The only panacea to hell is Love, which leads to God's Light, not to hell.

## MASTERS OF DARKNESS

That there is an order of high personalities in our Solar System operating as forces of darkness is not a new thought among men. It is given unto man to overcome the crafts of these forces and thus rise into the higher realms of Light, both individually and collectively. The Vedas, speak of Maya the 'negative Current of Power', on which the forces of darkness abide and by which they operate. In the Shariyat, this negative current of power that fights to hold man in subjugation, via multitudinous entities, is known as and called 'Kal'.

Siddhartha Gautama (the Shakyamuni) is famed for 'his victory over Mara', after which he became the Buddha, the 'Divinely Enlightened One'. Mara is said to be the leading master of the network of darkness in this Universe. Our Lord Jesus the Christ, the Blessed Immanuel, was even tempted by 'the Prince of Darkness' soon after His baptism by John the Baptist; and the Great Lord overcame. In the Bible the entity that tempted the Lord is called Satan, Old Serpent, Great Dragon, amongst others. (See Math. 4: 1-11; Rev 12:9). In the Koran (2:35) the 'evil one' is called Iblis. He is also called Satan:

> "O ye who believe! Come into submission wholly and follow not the footsteps of Satan; surely, he is your open enemy"
> (Koran 2:209).

The foregoing is a clear evidence that the people of old knew of the existence of the forces of darkness in one form or the other. But one thing is certain, from the point of High Metaphysics, the power that the forces of darkness have over man, is strictly based on the degree and the extent to which they are empowered and reinforced by man against man, by man against the world of man, as already exposed in this treatise.

## THE POINT THAT CONTROLS EVERYTHING

To this end, an Extraterrestrial Intelligence in a great unknown heavenly realm spoke to humans thus: "There are, then, no devils waiting to carry anyone off, unless you create (empower or reinforce) them yourself, in which case the power resides in you and not in the mock devils". Another thing we must understand is that nothing is independent and exist absolutely on its own. Even the so-called forces of darkness do not exist on their own completely. They exist, as all things in existence do, only on the One Eternal Universal Body of the Almighty God. Nothing can exist otherwise. As God said in the Bible:

> "I form the light, and create darkness: I make peace and create evil: I the LORD do all these things" (Isa. 45:7).

Eternal Validity of Oneness

Also Saint John wrote:

> "All things were made by Him; and without Him was not anything made that was made" (John 1:3).

In the Bhagavad-Gita (10:8) God puts it this way:

> "I am the source of all spiritual and material worlds. Everything emanates from Me. The wise who perfectly know this engage in My devotional service and worship Me with all their hearts."

In chapter 9 verse 6 of the Bhagavad-Gita, He said:

> "Understand that as the mighty wind, blowing everywhere, rest always in the sky, all created beings rest in Me".

When we study the Words of God in 'The Tiger's Fang', this point may become even more clarified. There (Chap. 2, p.30) we read:

> "Parts of the one whole idea are only seeming. There are no two separate parts of separable things in the universe. There is but one whole simulation of the one whole self, that is Myself - the idea. Every part of My worlds move in interdependent

unison as the wheels of a watch in unison. The wheels are geared mechanically together, and so the rhythmic waves of this world (Heaven) and all worlds below geared together electrically. The whole is Myself, and I must keep My Body balanced as one. Changes of conditions in any one part are simultaneously reflected in every other part, and are sequentially repeated in it, in cycles of waves flowing from My Heart to all My Universes, Galaxies and Worlds".

## THE ULTIMATE SOURCE

In human terms, GOD is like fire. In the first place GOD is the Source of Unlimited Multidimensional Energies, by which HE is Omnipresent everywhere at all times managing all things. We also see that fire is the source of three specific energies, namely, light, heat and smoke. If a room, for instance, is dark and you keep a burning fire in it's centre, the fire will become 'present' everywhere in that room. This is not directly on its aspect as fire, but via other energies arising from it. The light emanating from fire we consider to be good, because it dispels darkness and enables us to see.

However, this light is not the only thing coming from the fire. Both heat and smoke, each doing a different thing from that done by light, will also come out of it, all going beyond the specific location wherein dwells the fire. In the same way, all things emanate from GOD and are one in HIM, but divided in manifestation as opposites, as if irreconcilable. If a minor element like fire can produce energies that go beyond it's stationary location, we can at least from this have a clue that leads to greater awareness of THAT which we call GOD.

Consider light arising from fire, for instance, it has a fundamental duty that can never be performed by both heat and smoke. Neither can either heat or smoke perform the duty of light; yet they all have only one source of being, FIRE! Their unity therefore can not be within the framework of manifestation or appearance but solely within the system of their origin, their source. For further example: imagine great personalities arising from this light. As we hold in our mental evidence that light is good, we would consequently expect the personalities, forces, or entities emanating from it to function within the system of benevolence. To this end, they would become in our terms holy angels, saints, holy prophets, and all things associated with the sacred concept.

What of heat and smoke? We know that these have inherent within them the propensities of discomfort. With this understanding we would further consider the

Deeper Realities...

personalities, entities, or forces arising from them to be detrimental. These in human terms could then be seen as demons, witchcrafts, sorceries, terrorists and all things associated with the vicious concept. The actions of all those arising from light will, in the light of the foregoing, oppose the actions of all those arising from both heat and smoke. I am saying this by way of analogy.

Now, consider a situation in which fire (the source of light, smoke and heat) sets forth a decree that the personalities arising from light will not be ultimately defeated by personalities arising from both heat and smoke, or that the problems caused by those from heat or smoke can only be salvaged by those from light. This decree also states that the only basis of victory for those in light over those in heat and smoke is love. This becomes a fixed rule otherwise victory will elude them. Furthermore, this decree holds that light is the only channel or the only way through which anyone can return to the source, fire. This then entails that for anyone in heat or smoke to return to his source, he has to cross over to light and abide by the principles of its system.

## THE BEST PATH OF VICTORY

This analogy is all about God, man, the Forces of Light, the forces of darkness and all things in existence.

## Eternal Validity of Oneness

God is the Source of all things, but the rule which returns all things to the Source, which gives us victory over the systems of darkness, is Light, Love. The best way, the easiest way, to overcome the forces of darkness in all the realms, is to strictly abide by the admonitions of Lord Jesus the Christ, which says:

> "Ye have heard that it hath been said, Thou shalt love thy neighbour, and hate thine enemy. But I say unto you, Love your enemies, bless them that curse you, do good to them that hate you, and pray for them which despitefully use you, and persecute you" (Math. 5:43,44).

In 'The Everlasting Gospel', God spoke on this Love as follows:

> ".... If you love one another, you are walking in the light and there is no occasion of stumbling in you. Any person who loves is the person who is enlightened. He has seen very far. Any person who loves all is the person who has what surpasses the whole world" (TEG. VOL.1, 63:38, 39).

Any religion, marriage, family, community, business venture, government, or association not based on Love, is invalid. Any human life without Love is life in darkness; the life of such cannot reach God, cannot see God. When you possess Love, you radiate Light, you radiate God, then you are beyond all the subtle vicious forces of darkness.

Any sincere act, word, or thought of Divine Love performed in selfless service to humanity, is intercepted by the Forces of Light in the higher realms of existence, which they record as the 'work of God'. Such actions of Love as taking care of the needy, helping the less-privileged, praying for the good of others, providing for the down-trodden and advancing the defence of the poor and others, constitute the 'True Religion on Earth'. They are within the system of the Universal Brotherhood, which in this Age is set forth, via the 'Immutable Cosmic Laws', to play out supremely. They are such that serve as the basis through which the Holy Spirit will lift up our world.

## THE AGE OF WINNOWING

The earthmen have finally come into the Age of Universal Brotherhood, the Age of Earth Winnowing, the Age of the Paradise we seek, the Age of Spiritual Civilization based on Love, the Age of the final battle, the battle of all battles between the Forces of Light and the forces of darkness. In the Occidental Metaphysical parlance, it is called the 'Aquarian Age'. The 'New Age' began in our time on the first week of July, 1976. This was when Mars and Jupiter aligned with the Moon in the 'Seventh House of the Circle of the Sun', to characterize the highest peak of Divine Power on Earth, towards the reign of 'The Son of Man' - The King of Kings and Lord of Lords.

When we speak of 'Age' or 'Generation' from a Cosmic (Universal) point of view, it is based on the movement of our Sun around the Central Sun. In this movement, the Sun returns to the point of departure, or make a revolution, in about 26,000 years. The governing orbit of this movement is divided into twelve signets or signs. It takes the Sun and his family of Planets about 2,000 years to travel from one sign to another in the course of this movement. This period of about 2,000 years therefore is the measurement of 'Age' or 'Generation' in higher metaphysical or spiritual terms.

The Generation in which the Great Lord Jesus lived physically on Earth began in 168 B.C, at the end of the battle of Pydna in Macedonia, during which Rome overcame Greece. So when the Lord said,

> "Verily I say unto you, this generation shall not pass till all these things be fulfilled" (Math 24:32),

we should try to understand when the 'generation' He

spoke of began and when it ends, what to expect and what not to expect, amongst others. All these are in His terms, not in the mundane human terms, not in mere religious terms.

In this Age, the 'Kingdom of God on Earth' will be displayed with the 'Finger' of Supreme Light in the full consciousness of Universal Love. Nevertheless, a mighty conflict for the 'Soul of Earth' will come forth into the physical realm of Earth. The stage is set towards the full manifestation of this any moment from now. It will be the battle of all battles, a mortal combat, the final battle between the Forces of Light and the forces of darkness.

## CONFLICT OF POWERS

Now there is a high build up of armies on both sides, the prize of which is the entire human race. On the side of the Forces of Light is the 'Personification and Manifestation of the Supreme Light in Two Supreme Versions of HIMSELF'. This has highly empowered the Earth to give victory to the Forces of Light. These Two in manifestation are known in Bible terms as 'God and the Lamb'. They fulfill the ancient Bible prophecies, that in the course of time they 'shall' come within the surface of Planet Earth.

Eternal Validity of Oneness

Thus the Bible says:

> "And there shall be no more curse: but the throne of God and of the Lamb shall be in it; and His servants shall serve Him: And they shall see His face; and His name shall be in their foreheads"
>
> (Rev. 22:3,4.)

Also study the parable of Christ in Math. 21:33-43, wherein He coded this as the manifestation of the 'Owner of the Vineyard' and of course with His Son, as well as Rev 19:11-21 for His operation and that of the 'Armies' in the heavenly realms). We must understand that this has to do strictly with the Earth, the world of man.

## MAN AGAINST MAN

On the other hand, the expected victory of the Forces of Light is seriously being challenged by the earthmen, who continue day and night to strengthen the entire base of the mighty forces of darkness against the world of man, via their unchecked vicious mental pattern. If nothing is done now to raise the consciousness of man, to make him strengthened in the practice of Love, the wicked ones will have a strong base to manipulate things around by fixed 'Universal Laws'.

Love is and will always be the only way out for us. There is no other way for humans to triumph over the host of darkness in this coming conflict, except Love. Period! God Himself confirmed this in The Everlasting Gospel (Vol. 1,63:68) when He said to man:

> "... what will plague the world shortly cannot be subdued by any other thing except LOVE alone".

So the only problem remains the strength which the forces of darkness derive from humans against humans.

## CHILDISH NATURE OF EARTHMEN

At the moment, the earthmen are compared to about 500 children playing on one side of an open field. The forces of darkness are like 100 strong hungry lions on the other side of the field, who come forth to devour the little ones. The Forces of Light are like fifty well armed men from the 'Special Forces Unit of the Army', who came between the children and the lions, solely to annihilate the lions and rescue the kids. The Two Versions of the Holy Spirit, each which is the Mightiest of the Mighty, are like the commanders of the special forces.

The two commanders and their armed men, each having enough weapons to destroy all the hungry predators, stood between the kids and their worst enemies. But in their foolishness, the children were moving towards the lions to play with them. They thought that the lions had come to play. They also thought that the Special Forces standing between them and the lions were mere distractions, who wanted to do nothing but distract their fun. Thus, in their rage the kids began to mock and make jest of the armed men and their commanders, casting sand on them, yelling that they should get away so that their play will begin with the symbols of death.

The kids did and said all sorts of things against the special ones, unknown to them in their little minds that even their entire survival lie in the hands of the commanders of the armed men. If the two commanders become discouraged by the attitude of the foolish ones, and take a leave with their men, the kids will be destroyed. If they endure the kids and continue the battle for their rescue, the lions will be vanquished. This is a clear picture of how the people of this world are viewed now by the Wise Ones in the higher realms of Light. The people of this world are like a man whose house is on fire, but instead of working to quench this fire he engaged himself in chasing after rats. Consequently, when the earthmen finally learn the principle of true Love, or the brotherhood of life, it will surely come through a bitter lesson.

## AKASHIC RECORDS

In a message sent to the earthmen from a certain higher realm of Light, the realms of Akasha (from the 'Akashic Records'), through Dr. Levi H. Dowling (1844-1911), contained in the introduction of a holy book we know as 'The Aquarian Gospel of Jesus the Christ', man is further made to be aware of this. It was a revelation upon the cusp where Ages meet, the circle of the Sun governed by the 24 Ancient Ones, the 24 Elders, twelve Cherubims and twelve Seraphims; each of the twelve Ages ruled by two, a Cherubim and Seraphim.

It was a revelation that Dowling had on the transfer of the scepter of 'Dominion, of Might, of Wisdom and of Love' from the Guardian Spirits of the past Age to the Guardian Spirits of the present Age. From the reference point of his time then, he merely foresaw what would come to pass. At a point he took up his pen to write the things that were said, but the Cherubim of the outgoing

Age said to him:

> "Not now, my son, not now; but you may write it down for men when men have learned the sacred laws of Brotherhood, of Peace on Earth, goodwill to every living thing".

Furthermore, he wrote:

> "And then I heard the Aquarian Cherubim and Seraphim proclaim the Gospel of the coming Age, the Age of Wisdom, of the Son of Man".

His message from Akasha continues thus:

> "In the boundless blessedness of Love, the man was made the lord of protoplast, of earth, of plant, of beast.... and He who gave the lordship unto man declared that he must rule by Love. But men grew cruel and they lost their power to rule.... The coming Age will be an age of splendor and of Light, because it is the home age of the Holy Spirit; and the Holy Spirit will testify a new for Christ, the Logos of Eternal Love .... Man will fully regain his lost estate, his heritage; but he must do it in a conflict that cannot be told in words".

Deeper Realities...

## MESSENGERS OF LIGHT

In this Age we must be our brother's keeper. We must not sit idle thinking that because we are in the right standing with God, others can go to hell. We must understand that if our Planet shares in one way or the other in the activities of those in darkness, then we all will be part of it in one way or the other. It is our duty, those in Light, those who understand the potencies of the words, thoughts and actions of Love, to help others rise out of the gestalt of darkness.

By so doing we de-populate the human channels of darkness in the world of man. Through our constant selfless prayers and actions of Love, we will ensure that the forces of darkness do not have the base to bring forth highly perverted and most vicious humans to preside over the affairs of any nation, any state, any region, on Earth. The workers of Divine Love in the entire world of man, should strive to increase the rate of what they do for the sanity of our world.

# CHAPTER 9

## THE REALITY OF OUR TIME

Terrorism cannot be defeated on Earth by using hate to fight hate, but by using Love to banish the elements of hate. Even this in itself came into manifestation as a result of man's mass vicious mental pattern, so also is all forms of wickedness. Whatever happens in any nation, good or bad, is not by chance. Whatever happens to anyone does not come by accident, by luck, or by chance. Whatever kind of leadership that comes up in a nation at any given time, is the outcome of the inner mental mass dominating propensities of the people of that nation. We must continue the 'battle' until darkness in all ramifications is completely overtaken by Light.

In our human mundane limitations, we often see ourselves within the framework of the material body camouflage. To this end, we claim to be African, European, American, Arab and all that. We also identify ourselves often strictly within the system of a religion or tribe. We must understand that beyond all these, we are in our Higher Selves identified each as a soul personality. Even the Beings in the higher realms of God's Light do not see humans in these camouflage systems. From the higher realms, the earthmen are seen solely as humanity, the human race. This is regardless of tribe, religion, nationality, amongst others.

## NOTHING HAPPENS BY CHANCE

When a natural disaster occurs, which, as earlier stated, is the reciprocation of man's accumulated wickedness, it does not discriminate within it's area of manifestation. When an earthquake occurs for example in a specific area, it affects the people there, regardless of the religion, nationality, or tribe of anyone found there. Anything that happens to a person, a family, a community, a nation, or the entire Earth, whether good or bad, is based on what is in the psychic bank or the spiritual account of such a person, family, community, nation or the human race as a whole. There are constant debiting and crediting processes into all forms of personal and collective accounts of humans.

This is done in all the Universal Systems by diverse Versions of God manifestations in the Cosmic realms known as 'The Lords of Justice and Retribution'. And in their terms, nothing in all the Universes of God happens by accident. Thus, if you happen to be in a place where disaster occurs, you will escape or survive when there is a base for this in your psychic bank. But the very fact that you are there at that moment, or that you live or work there, is also not by chance. There is a subtle or metaphysical basis by which it came about. Consequently, even when you escape, the very basis by which you were there, will entail that you suffer some loss. This loss may or may not be in material terms. A wound, an emotional trauma, all constitute a loss in one way or the other. All things happen by the immutable process of God's fixed 'Laws', whether we know it or not.

## THE BIBLICAL LOT

Remember the incident of Lot as recorded in the Bible. There was some element of greed which made him chose to live in the place that was destroyed by fire. In the first place, he was not supposed to choose a land before Abraham, but after Abraham. In the human tradition, the elder takes a property first before the younger. Thus he failed a test and chose the land of inheritance first before his elder, Abraham. Even on that, he chose what in his terms was considered a better part, in which greed and selfishness became evident. He was far from what God told man in the Divine Iliad:

"Serve first thy brother. Hurt first thyself

rather than thy neighbor".

The place chosen by Lot finally led him to Sodom.

"....and Lot dwelled in the cities of the plain and pitched his tent toward Sodom" (Gen. 13:12).

While in Sodom, if he had consistently prayed and sincerely worked hard to raise ten human channels of God's Light, those channels would have served as the basis of some degree of Divine intervention for the entire people of Sodom and Gomorrah. Through them those cities would not have been destroyed. In the final prayer of Abraham to save Sodom and Gomorrah and the answer that the LORD gave him, it became clear that those cities did not have just ten people that could bring forth Divine Light to salvage them. (See Gen. 18:32).

Lot escaped the destruction of Sodom and Gomorrah but not without some loss. He lost his family house to the inferno. His wife died in the process of escape. He went and lived in a cave. All these were significant inconveniences to him, brought about by the mere fact that he lived in Sodom. When we understand the basis of things like this, we will see more reason to raise the stake of Light via the thoughts, words and actions of Love in our time.

## NOW IS THE TIME TO ACT

This is the time for us to fight very hard, to fight day and night, for the banner of Light to be raised over the forces of darkness in the field of battle. We have to work hard, pray very hard, for the increase of the human channels of Love in all the nations of this world. We must understand that the Astral vicious entities are working very hard to raise and sustain the human channels of darkness, which give them the base to do what they want in the world of man. They are not relenting their daily battle to invade the minds of the people of this world, so they could have enough to eat in their farm. It is said 'God forbid does not take away death'. To say: 'God forbid, the forces of darkness cannot do anything on Earth', is not enough. We have to fight now through Love to salvage this world.

We must understand the bona-fide weapon necessary for our use in this mighty battle. In the Bible, Saint Paul wrote:

> "For the weapons of our warfare are not carnal, but mighty through God to the pulling down of strong holds"
> (2Cor. 10:4).

Today many of us claim to fight against the network

*Deeper Realities...*

of darkness; but what is our weapon? What is our method? It is important we understand this, because we will fail if we use the wrong method.

## FIGHTING WITH THE WRONG WEAPON

In the 'Canterbury Tales' there is this story of three friends who took a strong collective decision to fight Satan and destroy him. In the course of their going around searching for Satan to destroy, they met an old man. When they informed the old man that they were in search of Satan to destroy, because of the untold problems and hardship caused by him everywhere on people, the old man told them that he saw Satan under an Oak tree a moment ago. They went to the Oak tree and saw a large chunk of gold.

The three friends began to thank God that, while searching for Satan to maim and destroy, they had finally stumbled on the fortune of their lives. Then they sat down and relaxed. One of them thereafter went out to buy food for all. As soon as he left, the other two plotted to murder him, while he also concluded to murder those two. Finally he poisoned their food; and when he came to them, they killed him. As soon as he died, the two began to fight and at last the stronger killed the weak. As the stronger settled down and ate the poisoned food he also died.

All of them died because each desired to possess the gold alone. The three friends who set out to destroy Satan were themselves destroyed by Satan through the lust for materialism. They set out to fight a warfare in which they did not possess the right weapon. Love is the sole weapon for victory over darkness. You cannot overcome the forces of darkness with lust, greed, anger, vanity and attachment to material things. These represent the 'great sea' which we must cross into the land of victory. In the holy book, 'Elucidations on Love', God said:

> "Two or three persons are better than a thousand, if only they have Love. There is no way any person can run a successful race, or be able to cross the sea, except he arms himself with Love".

## LOVERS OF HUMANITY

In a high metaphysical holy book known as 'The Twelve Blessings', a Mighty One of Light from a heavenly realm in the Venus Planet said:

> "Blessed are they who Love, for they are the Disciples of God.... These are the Ones who will save the pitiful ones, for

these are they who will become the very essence of the heartbeats of the pitiful ones, for they will be instrumental in helping greatly to transmute the only devil which exist upon this Earth. The Lovers of God, through man, will be the Ones who will light a Light in the hearts of all men, so that Wisdom may enter into these vessels so purified".

To live in Love is the only way to fight and banish the influence of darkness. To manifest Love is the only way to salvage our world; it is the only way to transmute the humans who the forces of darkness have captured as 'puppet machines' to advance the course of wickedness, which is their farm. When one is out to wage a 'war' against the host of darkness, he must have a proper inner safeguard, so that his mind will not come under the invasion of the same entities he is out to vanquish. This inner safeguard is 'Divine Love'. One must understand in the first place that the battle is in the mind, in the consciousness of one's mental pattern.

## ESCALATING INJUSTICE WITH INJUSTICE

Today in our world we see the total abasement of some humans who 'sincerely' set out to fight evil, to fight injustice, to fight the vicious manipulation of man against man, all in their terms. The enlightened minds can see how in most cases these humans are 'sincerely deceived'. This is because they lack the mental pattern of Love. Some of them have become suicide killers and core instruments of death, in the name of God and religion. Even when people have a just cause to fight for, in human terms, by operating within the system of hate they automatically come into the framework in which their minds, via the elements of reprehensive conditioning, become invaded by the most vicious Astral forces. These wicked entities in turn use such 'captured humans' to escalate in the world of man the same 'injustice' or 'evil' they had set out to redress.

The Astral forces of darkness use their human captives as mere robots or machines for vicious work. These captives work in this manner until they become free from darkness by the potency of Love, which operates the Light and Power of God. When man is transmuted from darkness to Light by Love, he begins to be assisted for ascent by the Forces of Light. He begins to live within the frequency of the Holy Spirit. He becomes alive in God's Creation, for one in the darkness of hate is already dead. The only way to cease to be a mere tool, robot, or machine in existence is to rise into the system of Divine Love.

Deeper Realities...

## WILLING TOOLS OF DARKNESS

The ancient sacred writings of Tibet known as 'The Shariyat Ki-Sugmad' (Book Two) has this to say about man:

> "He is a marionette pulled here and there by the invisible strings of the astral world. If he understands this he can learn more about himself and then, possibly, things may begin to change for him...Man is a machine, but a very peculiar machine. He is a machine which in the right circumstance and with the right treatment, can know he is a machine; and having fully realized this, he may find ways to cease to be a machine".

Some people claim to 'fight for the Almighty God' within their understanding of a religious system, but some dark forces are at work through them, thus they don't care about whosoever they murder in the course of their battle. They kill even those who belong to the same religion with them. Is this in itself not highly misguided? Is this not a clear evidence that they are not fighting for the very religion in which they claim to belong? For instance, when you take time to study the fundamental teachings of Islam, or when a non Muslim comes in close contact with a true Muslim, as I do, the display of sincerity, love and religious sanity will be evident in the Muslim.

In a book entitled 'The Way To Achieving Islamic Manures', by Dr. Ahmad Al-Mazyad and Dr. Adel Al-Shddy, we note these lines:

> "The more a Muslim adheres to the manners laid out by Islam in his daily practices, the closer he gets to achieving his desired perfectionism.... On the other hand, when a Muslim is distanced from Islamic manners and etiquettes, he is in fact drifting away from the very essence of Islam and its basic tenets, and has become like a robot which is devoid of a soul or any human feelings".

What of the churches of our time? There are some people among them who claim to be seriously out to banish the network of darkness. But in several occasions these are maimed by the instrument (materialism) of the very one they were out to cage. The Hindus and others are not left out of this game worked out by the craft of darkness. Humanity must turn fully to the path of Love and understand that this is the only true religion.

**Deeper Realities...**

The Grail Message specifically states that

> "After eliminating all distortions and dogmatic restrictions, the Religion of Love will be a doctrine of the strictest consistency, in which no weakness or illogical indulgence is to be found."

## MUNDANE RELIGIOUS PROPENSITY

We have this illustration of seven great clergymen from different religious denominations who gathered for a preaching engagement. The selected topic for all was 'The Love of God'. People also gathered to hear them. It was set out by the organizers of the event that any among them voted by the people as the 'Best Preacher of Love' will become the leader of the rest. The first clergy came forth and spoke. He declared that his church is the first among all, and that anyone who truly want to serve God in Love should come into it and be saved by 'The Son of God'.

The second clergy took his turn. He proclaimed total submission to God as the basis of manifesting His Love. He admonished the faithful to always Love God and be prepared at all times to kill or maim the enemies of God on Earth. The third clergy rose and proclaimed himself a 'Witness of God'. He categorically declared that his religious system is the best among men, the one and the only true denomination on Earth. He warned that anyone who wants to experience the Love of God in HIS Paradise on Earth must belong to his congregation.

The fourth clergy came out in style. He emphasized that anyone who claims to know better than him on the matter of the Almighty is a great liar. He made it known that he has experience beyond all in this matter. He said that he was born not merely once but another beyond the first, which puts him ahead of all. He warned that all men who want to go to heaven must abide by his understanding of this. He concluded that any Love shown outside his system of belief is not acceptable to God.

The fifth clergy stood and considered the rest as men of partial thought. He said that his religion wrote the first book about God on Earth. He claimed to have known it all beyond all others. Then he invited all men to come into his temple of worship to experience the Love of God in full. The sixth clergy educated all present on how to fight a spiritual battle. He told believers to Love God and show Love to the brethren. However, for the human agents of darkness, he called on the faithful to come upon the mountain and command the fire of God to consume such bad men.

## THE ONLY TRUE RELIGION

The seventh clergy was invited to speak, but he did not, rather he directed his drivers and other workers to begin offloading the goods he brought to the venue in twenty lorries. Then he began to call different categories of the poor to come out. To the hungry he gave food; to the sick he prayed for divine healing, gave some drugs and money; to the ones in rags he gave clothes, and much more. This he did not minding which religion one belonged. At the end he prayed for everyone, asking all to go and live in Love and peace. The people in turn voted him to lead all the clergies.

To help the poor merely to show off, without Divine Love in your heart, or what Saint Paul called 'charity' in the Bible, cannot bring you into the path of Light. But to do so in Love is the real work of God. Our Lord Jesus said that in the hour of fulfillment He would tell those who walked on the path of Love:

> "Come, ye blessed of my Father ... for I was hungered and ye gave me meat: I was thirsty, and ye gave me drink: I was a stranger, and ye took me in: Naked, and ye clothed me: I was sick, and ye visited me: I was in prison, and ye came unto me".

He further made it clear that whosoever does these to

any human, has done it to Him. (See Math 25:31-46) Earthman, was it not spoken of by the Wise Ones of old that showing Love to others is the true essence of religion? Remember the admonitions of Saint James in the Bible:

> "Pure religion and undefiled before God and the Father is this, to visit the fatherless and widows in their affliction, and to keep himself unspotted from the world" (James 1:27)

## THE PATH OF LIGHT

Herein lies a clue to the great message which brings one to the path of Light. It entails selfless service in Love, not only to the widows and orphans, but all humanitarian service in Divine Love are involved here. Whether this Love is demonstrated in thought, word, or action, it is what we need now than ever in the elevation of the human race. Therefore being a Pastor, Sheikh, Rabbi, Guru, Lama or Prophet alone without Love, does not make anyone a channel of God's Light on Earth. But living in Love makes you a channel of Light, even if you have no religious title.

When Love rules all life in the world of man, as it must certainly do soon, then the entire Earth will become the realm of Light, a Paradise for the children of men. As God said in the Divine Iliad:

> "There is naught but interchange of the Light of Love, for all creating bodies are centered by Me, and I am the Light of Love... Again I say, Love alone ruleth all things of heaven and earth. With Love I build My universe and with love it voideth itself in Me for again reappearing as My universe".

The first beneficiary of the thoughts, words, or actions of Love, is the person through whom they came forth, while the next beneficiaries include our Planet and the people on Earth. The first victim of vicious thoughts, words, or deeds, is the one who gave birth to such, while the Earth and the people therein share in one way or the other. The vicious thoughts, words, or actions of men have, in certain specific framework, caused mighty battles between the Forces of Light and forces of darkness. This is going on even now in the Astral zone.

## SEAL OF THE CHRIST

The Mighty Lord, Jesus the Christ, who remains the Divine Personification of Supreme Light, came on the physical realm of Earth to, among other things, head the battle of Light against darkness. He came as 'The Light of the World'. His Holy Blood that He deliberately shed in the course of this, including His resurrection, become the seal of victory for all humans who walk on the path of Light, Love. He did not come to form a religion, but to save the world by the seal of His Blood. He came to display Love meant as a guide for the children of men on how to work with God and for God.

The Lord came to lay the foundation of Light in the world of man. This opened the gate of power, mercy and life in Love on a higher dimension for man. This also became the foremost base for the empowerment of the Holy Spirit on humans, past and present, including the empowerment of the Forces of Light, for the battle is on to salvage Earth in our time. In this era the battle to banish darkness on Earth for Love to reign supreme has become far more serious than when Lord Jesus was physically on Earth. This is because the forces of darkness have made colossal advancement which completely made feeble what they were in the past.

On the other hand, the Mighty Forces of Light are not also folding their arms in the area of advancement. They fully understand the high level of preparation made in diverse realms of darkness for the final battle in this Age. To this end, the Forces of Light have on their side the Manifestation and Personification of the Divine Spirit in 'Two Supreme Versions' on the physical realm of Earth. These stand now as 'The Supreme Fountain' of the

> Deeper Realities...

endless Power of Light in all the realms of Earth and in all the realms of Cosmic manifestations. The only tool needed to stir the Holy Spirit into action, and for HIM to empower the Forces of Light the more for victory in the field of battle, is man's mental pattern of Love; Love in thought, word and action.

This is based on the 'Cosmic Law' that whoever needs help should ask for it. If man needs help from THAT which is above him, he must first ask the help of THAT. If man desires to be assisted by powers operating beyond and above the Earth, he must perform specific and necessary action on Earth that will, by 'GOD'S Immutable Laws', become the basis of the assistance he seeks, as well as the 'lawful base' of it's manifestation on Earth. Thus when man shows Love on Earth, he is knowingly or unknowingly projecting the lawful basis of help from above, from the One Eternal Holy Spirit, or HIS Forces of Light.

## MYSTERY OF OUR TIME

Some of these Forces of Light are here now; they are born in this Age as 'sons and daughters of men' on diverse ranks and levels in different nations. They serve quite clearly as the strongholds of the Spirit of Light in the world of man. Some of them in human terms have not come to the full realization of whom and what they are. But this does not stop the specific nature of work assigned to them by the Holy Spirit. Generally speaking they all live in the essence of high intensity of Love. From the point of the Universal Holy Spirit, it does not matter whether or otherwise the coming of these Mighty Ones are recognized in mundane religious terms. So also is the manifestation of the Highest Divine Spirit HIMSELF, The Mightiest of the Mighty.

The Holy Spirit to this end has, by any of HIS Supreme Personalities, directly carried out high profile forays into the highest chambers of the Astral vicious denizens in our time. The Holy Spirit has also done this from time to time, in our time, via some mighty personalities in the order of the Forces of Light, whether those in the physical body or those outside of it. The actions of Love performed by men on Earth have always contributed to the success recorded in the course of these attacks against the realms of darkness, whether we know it or not.

## WAR IS ON

As there are some humans who serve as the strongholds of the Spirit of Light on Earth, there are also some humans who serve as the strongholds of the gestalt

Deeper Realities...

of darkness in the world of man. The war is on! The great battle going on in the realms beyond now, cannot be practically known merely by attending a religious service or by reading the holy books. While these are important in themselves, the practical aspect of invisible warfare comes into a complete different framework of experience. This also includes the outcome of each and every attack of the Forces of Light into the realms of the forces of darkness. The problem with some aspects of man's organized religion is that it claims that what it knows, the little it knows, is the highest of all, beyond which there is nothing to be known.

I have always said that the holy books are like signboards. A signboard merely points to what it represents. For instance, the Holy Bible speaks of God, but the Bible is not God. By reading the Bible we hear about God and by hearing we could develop faith to believe in God. But we cannot see God by reading the Bible, we cannot experience God in practical terms by hearing only if we do not have God, Love. An eatery called 'Fresh Restaurant', for example, could have its signboard some distance away pointing to it. If you go to the signboard and limit yourself by it, not taking any step beyond it, claiming it to be the eatery because you see on the signboard written 'Fresh Restaurant,' then you are not wise.

By limiting your movement to the signboard, you will not have the food from that eatery point. But by going beyond the letters of the signboard, knowing them to be clues that direct people to the restaurant, you will reach your destination and enjoy what you would not have enjoyed by remaining with the signboard. Job said about God:

> "I have heard of thee by the hearing of the ear: but now my eyes seeth thee" (Job 42:5).

If we fail to come to this platform in the present life, then we limit ourselves by reading and hearing of those who did.

Remember, however, that if all you know about the Source of your life (God) is only what you hear and read about others, or from others, then you know nothing as you should know. The life and works of those we read of, who while on Earth did experience God in the higher essence of Light and Love, past and present, serve as evidence that such is possible. For what one man is, another man is. The difference is on the intensity of experience, realization and awareness. Now, apart from seeing the Lord in His Personalization and Manifestation in any realm, we also see Him via the practice of Love, which He is, by the purity of heart devoid of hate:

> "Blessed are the pure in heart: for they shall see God" (Math 5:8).

Deeper Realities...

# CHAPTER 10

## BATTLES TO SUSTAIN OUR WORLD

The earthman on a general level do not understand nor appreciate the magnitude of battles constantly waged by the Forces of Light against the realms of the forces of darkness to sustain our world for GOD in Love. The earthman does not know the great work done in the realms beyond to salvage Planet Earth from total annihilation by the vicious network of darkness. In this regard, it can be rightly said that man is sleeping while his house is on fire.

## GAINING GROUNDS OF VICTORY

Do you know how the disintegration of Soviet Union came about? Do you know why America and Russia sat at last before a 'table of peace' and began to work with others for the prevention of the use of atomic bombs? These and many others came into physical manifestation as a result of the victory gained by the Forces of Light in the battle for the sanity of our world.

In a well coordinated foray into a high chamber of darkness, the Personality of the Divine Spirit with five high ranking Forces of Light came upon the base of a mighty Astral entity.

It was this entity who masterminded the invention of atomic and hydrogen bombs on the physical realm of Earth, via mundane scientists. The vicious entity was captured that same hour. The entity was later transmogrified whilst the transmogrified one was again banished from this Solar system. In the course of time, man will fully understand that without the Personification and Manifestation of the Holy Spirit in our time, this world would have perished.

In another foray into a well organized chamber of the forces of darkness, the Personality of the Holy Spirit stood upon a high base of the occult. This is the psychic base specifically set up by the hierarchy of the Astral forces to coordinate and ensure that ubiquitous atomic war is fought on Earth. This base was completely dismembered and put out of use. The mighty Astral entities there, forty eight in number, were all arrested, transmogrified and taken away. Even the very Astral Being who coordinated the crucifixion of Jesus the Christ is now under chains. He is as good as dead now in human terms.

Deeper Realities...

## REGROUPING OF FORCES

Though there are lots of high profile re-groupings in different Astral realms of darkness for the final battle to take over the Earth, which comes up any moment from now, the Forces of Light are also strengthened more than ever for this final battle. While the waiting game for this battle of all battles in the realms beyond, for the 'Soul of Earth', continues on both sides, there are occasional forays into some zones of darkness by the Forces of Light to arrest and transmute some vicious entities that come up, as well as to emancipate humans in captivity. This final battle is fixed to hold when, in human terms, the Earth Planet will vomit.

The forces of darkness also make occasional foray on the physical realm of Earth, via willing human channels which they have many, to inflict vicious pain and take more captives. In Nigeria, for instance, it is on record that armed robbers on more than one occasion while robbing travelers in the night on the highway, forced them to lie flat on the road. These innocent ones were crushed in large number, some beyond recognition, by other oncoming vehicles on speed. Things like this, or related to it, happen from time to time in different parts of this world. They are manifestations of the forays carried out by the forces of darkness on the physical realm of Earth, which in human terms is like 'guerrilla warfare attack'. The destruction of the World Trade

Centre that took several lives in New York, the mass murder of millions of Jews by Hitler, the ethnic genocide in Rwanda and other places, the ethno-religious massacre and reprisal massacre in Nigeria, terrorism, among a host of others in diverse places, are within the forays of the vicious Astral forces.

The main battle is in the realms beyond the physical, which determines the aforementioned manifestations in the world of man. These manifestations as a rule must occur via the human channels, whose mental patterns serve the interest of the forces of darkness. On the other hand, the Strong Spirit of every highly empowered channel of Light is involved in the war to enthrone Love over darkness on Earth. Some of these are physically manifested as humans with the aforementioned Supreme Manifestations, while others are not on the physical realm. But they are all doing the same thing as I earlier stated. The ones on Earth serve as the base, the reference point, for the invisible ones to act. While the invisible ones also reinforce those on the physical plane to act; all done strictly by the Holy Spirit, because, as the Bible says: "Iron sharpeneth iron" (Pro. 27:17).

## MY PERSONAL EXPERIENCE

In one occasion, I was caught away in spirit into a high realm of Light. The Two Versions of the Supreme

Deeper Realities...

Personifications of the Holy Spirit in our time came forth. I was directed to lead a battle into a particular zone of the realms of darkness for the arrest of a vicious one. This was the entity incharge in that zone of capturing humans for ritual sacrifice on Earth. The foray I made into this zone brought me face to face with this vicious one. He was a 'complete mad' Astral entity traveling upon the wind. On his left hand were seven strongholds, the instruments of captivity. On his right hand was a short rod, which I immediately understood to be the signet of authority of the vicious Astral forces.

As soon as he saw me, he came with a great speed for attack. I stood in a spot till he came. The rays of Light which came forth through me completely discomfited him. He stood there before me bamboozled. He could not act or go further. I commanded him to surrender his signet and strongholds, which he obeyed beyond his resistance. Before I went for this arrest, I was specifically told: 'Beware of the game that would be played out'. When this entity was disarmed, a stronger entity appeared in the form of an old woman. She congratulated me for the victory and requested that I hand over the defeated one to her for necessary punishment. I thought she was on the side of Light; and I was about to yield to her request when I consequently got a Divine message that she came from a higher realm of darkness.

This old woman was in-charge, in that zone where the other entity served, of the potency of blood ritual in the accumulation of money by wicked humans. I hit her forehead once and she fell down. Both of them were transmogrified. The Forces of Light from other realms later took them away. There are still many of their types in different Astral zones of the realms of darkness.

In a certain zone of darkness, there is an entity who joins forces with his kind to, among other things, get hold on children right from the womb for the works of darkness on Earth. The ones they had power over include those who in the course of the sexual act that brought them into the womb, the male or female involved was either drunk, had grudge against anyone, or was exasperated not less than three hours before the sexual act. When I made a foray into a realm of darkness for the capture of one of these, I saw him before he saw me.

He threw two Astral darts at me, but I held them as a piece of foam. He escaped into the invisible and I followed him up till he was captured. These things are too numerous to state here. It is a battle in which all the channels of Light are involved in one way or the other. Even the once evil forest in my village was destroyed in this manner. I am using my personal experience in this as a reference point. Every channel of Light has his. The evil forest was known as 'Ohia Ogwugwu' or the 'forest of Ogwugwu'. This Ogwugwu is a power of darkness to which humans were sacrificed. The Forces of Light came as little ones from above; a little one who came from my stock joined me with them in a foray made in spirit against Ogwugwu. He was arrested through the water we used which became fire. The little ones from above took him away in chains. Five weeks later a physical fight came up between youths for the sake of the forest. The government came in. Today, the evil forest is no more.

## WINNOWING OF EARTH

I have said in the course of this treatise that there was a time when Earth as a living conscious Being was sick and about to die, which would have ended all life forms on this Planet. I have also made it clear that this came as a result of all forms of hatred and wickedness practiced on Earth by humans. It was to salvage the Earth from death that the Holy Spirit came into manifestation.

Earthman, have you seen any tree that has its roots upward and the branches downward? Such does not exist in the absolute sense, but it does exist in a relative sense. If you go to the bank of a river and see a tree, you will see it stand in the absolute sense like any other normal tree. But when you look at the reflection of that same tree inside water, it would in a relative sense be turned upside down; that is, the branches will be down while the roots will be up.

I am speaking in this connection to our Earth. In a relative sense, the physical realm of Earth and all life therein reflect the existence of other realms. These include the metaphysical and spiritual realms. The life of Earth, the life of every creature on Earth, including the life of man, the highest on Earth, are manifestations of the One Eternal Spirit. These manifestations occur within the framework of the twenty four material and subtle elements that serve as channels of the life-force in the Cosmic manifestations.

## MANIFESTATIONS OF GOD

These primordial elements arise from hidden sources beyond what is known in human terms. The Sole Ultimate Source of all sources is the Holy Spirit. In other words, the invisible worlds which include the metaphysical and spiritual worlds are on different unlimited levels. The Sole Ultimate realm of all the realms is the core realm of the Holy Spirit, the realm in which nothing can be said of, the realm beyond all manifestations.

> Deeper Realities...

For the Holy Spirit therefore to come into manifestation in any realm of existence, as a Personality seen by Angels, Archangels, Cherubims, humans, amongst others, is the consummation of the mystery of all mysteries. It must be understood that each Personification and Manifestation of the Holy Spirit, 'The Absolute Spirit', in any plane of the entire Cosmic realms, is 'God in Manifestation'. The manifestations of God are unlimited in eternity. All the manifestations of God, whether as a Dove, a Lamb with Seven Eyes and Seven horns, a Rock, a Man, or in forms completely unknown, consummate what we know as 'The One Eternal God'. This is so because They All operate with One Love, One Eternal Light, One Purpose, One Desire, One Action, One Law and One Mind. Thus, GOD forever is ONE!

Each of the Two Manifestations of the Holy Spirit in our time has other Versions, several Versions (Personalities) of HIMSELF on different levels, also in manifestation in the Cosmic realms, some even on Earth. These multifarious Versions and the Great Ones of Light already in the Cosmic realms for ages before their manifestations, constitute the 'Forces of Light'. All the Forces of Light are coordinated by the Two Supreme Personalities, or what the Bible call 'God and the Lamb', now in manifestation.

## SUPREME PLAN TO SALVAGE EARTH

That all these are taking place now even on the physical realm of Earth, show the highest dimension of seriousness in motion to salvage Planet Earth and prevent a repeat of the past. In our Solar System a Planet was destroyed by the forces of darkness. The point we are now was the point from which that Planet was destroyed. The destruction of a Planet is a big problem in any Solar System. The situation in the earthly days of Our Lord Jesus is a child's play to the situation we are now. The Personification and Manifestation of the Holy Spirit in Two Supreme Versions at once is occurring for the first time, at that level, in any Planet of our Solar System. It has not happened before.

This is the highest demonstration of Love by GOD for any Planetary existence. It underscores the seriousness for salvaging Earth from the high powered forces of darkness that encompassed her. Without this highest dimension of intervention, a Planet will certainly be destroyed again in the Solar System, Earth! The Supreme Divine Light that came forth to help our Planet, came solely in the Spirit of Eternal Love. In manifestation, HE became the Highest in all Systems. In the Galaxy HE came forth as the LORD of the Galactic Lords. In the Solar System HE stood (The Two Versions) as the LORD of the Solar Lords. The Mighty Ones above recognized HIM as the Holiest of all the Holies.

Deeper Realities...

When the Planet Earth was about to die in less than fifty five years ago, the Holiest of all the Holies and five of the Mighty Ones of Light held the Planet, but not in human terms. And for the total healing of Earth, a primordial substance of the One Eternal Spirit was injected into her, for the greatest radiation of Love. The Holy Spirit salvaged the Planet Earth by this extraordinary means.

Then the living Being we know as Earth came forth in power, having received 'Divine Healing' from the great illness, a burden that came upon her via the long accumulated practice of hate and wickedness of the earthmen. The wickedness which culminated in the manifestation of atomic and hydrogen bombs on the physical realm of Earth. These bombs completely set the final stage for the death of Earth, the greatest calamity in any Solar System.

CHAPTER 11

FRAMEWORK FOR A NEW WORLD

The 'Primordial Divine Substance' administered by the Holy Spirit on Earth, did not only heal our Planet as earlier stated, but also infused in her multitudinous and multidimensional 'Rays of Divine Love'. This was consequently stored in the ultimate inner core of Earth, a sacred chamber guarded by Divine Flames of the Almighty. This high substance of Divine Love is programmed by the Holy Spirit to transform Earth, that the children of men (and every creature) will, after a short while, live upon her for good, only on the basis of the practice of Love.

HIGHER EXISTENCE IN LIGHT

At that time, the Holiest of the Holies stood in the spirit realm upon the Asteroid Belt, wherein a Planet once died in our Solar System. There HE made an eternal decree upon Earth. The Supreme Decree holds that: 'Atomic war will never come upon Earth, will

> Deeper Realities...

never be used to destroy Planet Earth, for Love is in full control of Earth'. Thereafter the Mightiest of the Mighty commanded the 'High Spirit of Divine Love' to bring about the necessary condition for the higher existence of humans in Light.

Herein lies the existence which fulfills the ancient prophecies of the 'coming of God's Kingdom on Earth'. The existence of the consummate consciousness of the brotherhood of all life, which constitutes 'The Paradise' we seek in the world of man. To this end, the aforementioned 'Primordial Divine Substance', which the Holy Spirit administered on Earth, was further designed by HIM to, after a short while, cause the Planet Earth to vomit in human terms. The hour of this in human terms is very near.

At that time our Earth will surely vomit out, in our human way of speaking, all the elements, instruments and human channels of darkness opposed to the reign of Love; for Love alone will dominate our Earth. At that time also the final battle of the Forces of Light and the forces of darkness will be staged. The gas pressures of Planet Earth will be released to bring forth great destruction, as two minor Planets will come out of orbit in our Solar System to strike one another. Some great mountains of Earth will become flat lands; rivers will come forth in the desert; some flat lands will rise as mountains and high hills, amongst others.

## EXPECTED CHANGES

Then shall the children of men bear witness of the eternal potency of Love. The mystery of Divine Love will be played out. No true human channel of Love in the world of man will be consumed by the great holocaust, the fire which will ravage even the sea. Then shall man know that Love is the only true religion on Earth. Then shall many recognize that the things spoken of our time by God via the Wise Men, Prophets and Sages of old, will in fulfillment be in God's terms, not in the terms in which humans understand or interpret them.

As the earthman comes into the new world of Love, he comes into the system of great change. New areas will be activated in his brain which will take care of the changes that would occur. Already it is a scientific fact that the earthman has not used more than ten percent of his brain power. Some calculate even less than this. A scientist, Sir Arthur Clark, in his 'Frontiers of Consciousness,' wrote:

> "Probably 99% of human ability has been wholly wasted; even today, those of us who consider ourselves cultured and educated operate for most of our time as automatic machines and glimpse the profounder resources of our minds only once or twice in a lifetime".

Deeper Realities...

Several changes will soon come upon the earthman on the physical realm of Earth. For instance, man will communicate with man, no matter the distance, through mind, as telephone will be of no use; there will be no labour pain in the course of child birth; hate and wickedness will not exist; man will move with the power of his spirit to any place without a car or airplane; there will be 'One Fold' under 'One Shepherd', among a host of others.

## THE FUTURE MAN

What man will become on Earth may sound bizarre from the point of what he is now. In the same way, what man is now was unbelievable to the people of old. Some people in the past foresaw the scientific inventions of our time, which were considered impossible by the people of their time. Mother Shipton of Scotland (1488-1561) was one of these. In her time she foresaw some of the inventions of our time.

The book, 'Collections of Prophecies,' records the fulfillment of her prognostications as follows:

> "A time shall come when carriages without horses shall go (AUTOMOBILES);under water men

shall walk (SUBMARINE); in the air men shall be seen (AIRPLANE); around the world thought shall fly (TELEGRAPH/TELEPHONE); iron in the water shall float (STEAMSHIP)... Then shall come 'The Son of Man'.... With a number of people shall He pass many waters, and there shall be battle among many kingdoms; and there shall be peace over the world; and there shall be plenty of fruit; and then shall He go to the land of the cross".

In the days ahead of us, man will have a sane world. He will practically understand his relationship with Earth and possess a high dimension of freedom in Love which did not exist in the past. He will fully comprehend the mystery of the Personification and Manifestation of the Holy Spirit in His Supreme Version as 'The King of Kings and Lord of Lords'. In the course of time all pregnancies will occur without sexual act.

In the course of time also there will be no death on Earth in human terms; for the psychic gulf between 'the living and the dead' will be severed; the 'living' will know the 'dead' and the 'dead' shall walk among the 'living'. The consciousness of man will experience 'A DAY' of the LOVE of the One Eternal King of Kings, who brings forth the conditions of Love and Peace in which the earthman will flourish; and this 'DAY' in our terms is 'eternal'. There will be no literal mundane proclamation of the King of Kings and Lord of Lords in the sky of the physical realm of Earth, for the earthman to kowtow, but the radiation of HIS Love separates the 'sheep' from the 'goat'. The future man will be a 'spirit' from the point of the present mundane man.

Then all institutions, governments, religious groups and churches will have a new beginning in Him. The Earth will be completely purged of hate. Joy will abound beyond the sea. The young will know that which belongs to the old. All the problems of man associated with spiritual ignorance will cease. These and several other changes will come to man on the physical realm of Earth in this Age. From the point of the present mundane man, it would seem as if humanity have changed into something else.

## EARTH IN THE WORLD OF STARS

One thing is sure in this Age, the affairs of men will not continue on the mundane path. Love will certainly rise from the hidden chamber of Earth, by the Finger of 'The Shepherd of Shepherds', to overwhelm all the realms of our world. By this, the Planet Earth will possess her rightful place in the world of the Stars. And now the Planet Earth do rejoice, for the King of Kings and Lord of Lords with the host of the Mighty Forces of Light are upon her. She rejoices because this is the time of God in her time, the hour of fulfillment in which she must rise.

Our Planet is in great joy because the SEAL of the Mightiest of the Mighty has descended in her midst. The joy of our Planet emanates from the knowledge that the hour is here when every eye shall see 'The Supreme Light' upon her. Earth do rejoice again and again that the greatest danger upon her is taken away whilst she will no longer be destroyed. She knows that the Holiest of the Holies has restored her rightful place of existence and that she will hereafter come forth and shine like the Stars in the realm of Heavenly Lights.

Therefore I say: Earth rejoice, for it will surely come to be that anyone who lives in Divine Love, will rise with you into your new place, the refuge of a new world. Anyone who continues to live in the framework of the darkness of hate, wickedness, vicious mental pattern and terrorism, will surely crumble in this Age. In the course of time such ones cannot stand the high vibration of great Love and Light which the manifestation of the Holy Spirit has already started to radiate on Earth.

Deeper Realities...

## PLAY YOUR PART NOW

Earthmen, this 'Great Light of Love' is radiating step by step in the world of man. The hour comes when it will overwhelm in the highest velocity all the realms of Earth. Do not pass over lightly this message of Love; for you now live, O man, in the era of the 'Last Covenant' for the entire creation on Earth. This sets the stage of 'A NEW EARTH' based solely on Love. By the Finger of the Holiest of the Holies, by the 'Signs and Signet' of 'The Son of Man', the 'Last Covenant' of a 'New Earth' has taken root in our Solar System.

Earthman, rise even a little above your daily struggles for mundane survival, take a little step above the quest for the possession of the temporal material things of life, observe and you will recognize the primary evidence of a New Earth. This is manifesting in diverse minute forms in the world of man. To firmly uphold this and overcome the elements of darkness, we must seek the Light, live in the Light by upholding the consciousness of Love in all ramifications. Therefore I say to man: Whatever is your religious denomination, whether you are old, young, rich or poor, male or female, black or white, live in Love with one another.

Stand firm, you human channels of Light; let your daily prayers to the Almighty come forth from the chambers of Love. Give the potency of your daily prayer to God, your time, the energy of your life, even the material things you have, for the service of Divine Love to others. This is the work of God.

Never you at any time of prayer command the 'Fire of the Almighty God' to consume any human, who you know or suspect to be your enemy. You must understand the serious Divine implications of this now within the framework of the 'Last Covenant' for all the earthmen. For some time now I have been warning against this. It is another way of feeding the host of darkness, because it is a prayer that comes from a wrong spirit, the vicious spirit of hate.

## PSYCHIC TERRORISM

There is no place in the Scriptures wherein Lord Jesus the Christ directed that this should be done. When you command fire to consume your fellow man, you become the first victim of this, via the 'Law of Reciprocal Action'. In those days, concerning a village in Samaria that opposed the Lord, the following is on record:

> "And when His disciples James and John saw this, they said, Lord, wilt thou that we command fire to come down from heaven, and consume them even as Elias did? But He turned, and rebuked them, and said, ye know not what manner of spirit ye are of. For the Son of man is not come to destroy

men's lives, but to save them. And they went to another village" (Luke 9:54-56)

. It is important that we understand the spiritual and metaphysical significance of not making prayer based on the core practice of hate.

Today in some religious gathering, the earthmen who are in the position to lead others, to provide guidance on the path of Light, give the multitudes stones instead of bread. They direct the congregation to lift their thoughts and command the fire of Heaven (the Holy Ghost or Holy Spirit Fire) to come forth and consume their human enemies in the world of man. When the reciprocal effect of this vicious mental pattern, in the name of prayer, begins to work, the fire starts from the bosom of the initiators. Such initiators are elements of psychic terrorism.

The vicious mental pattern that brings forth any form of prayer, not situated in Love, runs against the Divine Love in which lies the victory of Light over darkness. Those involved in this may not know that their aura and psychic systems are turned into the elements necessary for the aggrandizement of the most vicious Astral entities. They serve on a certain degree as the human channels of the forces of darkness, whether they know it or not.

For example, there is this book (name withhold) of 534 pages, a well known international Christian book that claims to educate man on how to defeat his enemies. In page 64 of this book the 'believer' is directed to recite this prayer:

> "You my uncompromising enemies, I decree that confusion will fall upon you now. Begin to oppress yourselves without a break. See yourself as your worst enemy. Block all your ways of blessings with your own hands. I decree that you will be so confused to the extent that you will stripe yourselves naked before the public. Begin to behave abnormal before your own helpers".

## CHAPTER 12

## WALKING ON THE VICIOUS PATH

It is clear that the prayer in the previous chapter, which from the point of Universal Love is another form of 'witchcraft', is directed against some humans expected by the author to go "naked before the public". This is among the worst form of hate. It is important to note that perverted messages and books that encourage this vicious system of prayer exist in large number today in the organized Christian religion. It should be understood that this represents an infiltration of the network of the vicious Astral entities into the most holy teachings of Our Jesus the Christ.

Leading people along this vicious path in the guise of religion and spiritual warfare, is certainly like the blind leading the blind. The Lord warned:

> "Let them alone: they be blind leaders of the blind. And if the blind lead the blind, both shall fall into the ditch"
> (Math 15:14)

## THOUGHTS POTENCY

That a vicious kind of prayer appeals to the mundane sentiments of people, on how they would want their human enemies to be dealt with by THAT which is above, can never grant such prayer entry beyond the circle of the macrocosm into the Divine Realm of the Holy Spirit. We must understand that 'prayer-thoughts' are valid things, as thoughts are things. Siddhartha Gautama (Buddha) said in the Lotus Sutra:

> "We are what we think. All that we are arise with our thoughts.

The wisest man of his time, King Solomon of ancient Israel, wrote in the Bible that as the earthman "thinketh in his heart, so is he" (Pro 23:7). It is said that Love attracts Love and hate attracts hate. Thoughts operate strictly under the 'Law of Homogeneity', from which you rise or sink according to your mental pattern. If you continue on the wrong path of wicked prayers, you will surely be the first to reap from them as a matter of must. There is a saying that 'the gods work in the farm of those who work in theirs'. Another saying has it that 'the stone that lies on the bed of the river should not complain of feeling cold'.

Richard Carswell in his metaphysical book entitled 'The Law of Attraction' wrote:

Deeper Realities...

> "Scientific research has documented what philosophers have known all along that negative thoughts produce the release of stress hormones that cause negative feelings. Positive, optimistic, joyful thoughts cause the release of endorphins and other beneficial brain chemicals that cause good feelings".

## THE MAKING OF A NEW MAN

If you engage yourself in vicious prayers to the Almighty, asking HIM to send fire upon your human enemies, it will by the 'Universal Law of Homogeneity' attract the things of its kind and bring these on its way back to the starting point, the originator, yourself! Then you will have more vicious things to contend with in diverse aspects of your life. To this end,

The Grail Message tells man:

> "Pressure and condensation produce the quality of a magnetic activity, in accordance with the Law that all that is stronger attracts what is weak. Similar

thought-forms are thus attracted from all sides and retained, constantly reinforcing the power of your own, your original thought ..... Let us return to the thought that attracted the other forms, and thereby became strong and even stronger. It finally emerges beyond you in firmly united power-waves, breaks through your own personal aura, and exerts an influence upon your wider environments".

God said to man in the Divine Iliad:

"And now shall a new unfoldment come to man who shall arise from the clay of him into My Kingdom of Light in My high heavens. I, the Father of man, await man's ending of his own agony, for man hath not yet known Rest, nor Love for he hath not yet known Me".

## THE REALM OF UNIVERSAL FREEDOM

Earthman, know and understand that there are three kinds of prayer, out of which one is the best. The first kind of prayer is the true prayer that comes from the consciousness of Universal Love. It is the prayer based on the transformation of wicked humans; a prayer of Love for the healing of the sick; a selfless prayer for the Light of the Holy Spirit to descend upon all men on Earth, and much more, within the framework of Love.

This kind of prayer is such that goes faster into the 'Divine Point' that coordinates answer to prayers, which we know in religious terms as 'The throne of God', and this Divine Point is present everywhere including where you are now. Selfless prayer is such which comes within the system of the 'Last Covenant', in the strength of the consciousness of the brotherhood of all existence. It is like a man traveling by flight, not by road, from Lagos to Accra. And by 'Law' the answer point of this kind of prayer is the person that originates it. Thus, in answer to all forms of selfless prayer, your personal need, as well as the battle of the vicious humans against you, will come into foremost consideration.

Through selfless prayer you serve others first, yet in this you are put first by THAT which is above. Christ said in the Bible:

> "And if ye have not been faithful in that which is another man's, who shall give you that which is your own?"
> (Luke 16:12).

## Walking on the Vicious Path

In other words, if you are faithful in selfless prayer for the advancement, goodwill and the like for other humans, yours will surely come into manifestation. You don't get vengeance by asking God to consume your fellow man with fire. If you follow the path of Love towards your so-called human enemies, the willing ones will thus be transformed, whilst the most vicious ones, who continue on the path of wickedness against you, will be dealt with by 'God's Immutable Laws', through the Higher Forces assigned to such.

In this connection, God said in the Bible:

> "To me belongeth vengeance, and recompense" (Deut. 32:35).

> "Thou shalt not avenge, nor bear any grudge against the children of thy people, but thou shalt Love thy neighbour as thyself: I am the LORD" (Lev: 19:18).

Saint Paul also wrote:

> "Bless them which curse you: bless, and curse not" (Rom.12:14).

Through selfless prayers, you are put in the realm of universal freedom.

Deeper Realities...

## TAKE LESS GIVE MORE

The second kind of prayer is selfish prayer: 'O God make me rich; O God make my football team to win this match; O God make me famous; O God give me this, O God give me that', amongst others. That is selfish prayer. This is not completely bad, because in one sense it demonstrates that you are expressing your confidence and dependence on God. However, this is like a man traveling on road with a lorry from Lagos to Accra.

Even if you get what you want through this kind of prayer, it must be understood that if you do not balance it with some elements of selfless prayers, then you are taking from, and not contributing to, the universal pool of Light and Love. If you continue on the path of selfish prayers at all times, you slow down the progress of your rise into the higher realms of Light. Even often some of these prayers are discarded by THAT which is above. And when you pray thus, 'O God where are you', or 'if truly you are God, come and do this or that', it becomes void.

## CONTRIBUTING TO THE POOL OF DARKNESS

Now, the third kind of prayer is the vicious one. The prayer that seeks the harm or destruction of a fellow man because he is your enemy, is a vicious prayer. This has no place in the realm of Light. If you continue in this (unknown to you), you become a major contributor to the pool of Astral power from which the forces of darkness draw their instrument of manipulation.

Apart from giving strength to the forces of darkness to act through vicious prayers, the 'Law of Reciprocal Action' ensures that the originator shares first from the fruits of this. Vicious prayer belongs to a mental pattern, a 'manner of spirit' that upholds the element of the modus operandi in which darkness thrives in the world of man. As Christ said to those on this path, "Ye know not what manner of spirit ye are of". That 'manner of spirit' is surely that of darkness that begets darkness.

Through vicious prayers you become like a stone that lies on the bed of the river. This stone should not complain of feeling cold, it should bear it because that is its lot. Therefore a man of vicious prayers should be prepared to bear the outcome of this upon himself. A man on this path will learn bitter lessons till he makes a change for the better.

Deeper Realities...

A woman from this path once came to see me from Kano when I visited Lagos. The meeting was by appointment.

## HINDERING YOUR RISE IN LIGHT

The woman believed that she was under the attack of the forces of darkness. As she sat before me, I could see quite clearly (with the eyes which transcends the mundane) the specific imprints of hate vibrating in her aura. She was a 'Born Again' Christian in her terms. Her tale held that her boss, the Area General Manager of a bank, was the human agent used to attack her by the forces of darkness. The AGM had officially issued her a query on a certain issue. She claimed that the man hates her with passion and that, as a member of a secret cult, he dislikes 'Born Again' Christians.

While seeking for solution over her imponderable situation, she met with the General Overseer of her church, a 'prominent' one in Nigeria. The 'Man of God' blessed olive oil for her. He directed her to be in the office before the AGM every morning for seven days, anoint his office door, say a prayer commanding the 'Holy Ghost Fire' to consume 'the agent of darkness and his plan', and be on fasting for the seven days.

After doing all these with intense prayer and fasting for seven days, the result that came forth got her completely bamboozled. Instead of getting any solution, the whole situation escalated: She was placed under suspension arising from the query. However, by mercy and for her transformation, a Divine message came forth from the higher realms of Light, the Holy Spirit.

Consequent upon this, I directed her to completely forgive the AGM, send him thoughts of Love and goodwill, fast for twenty four hours (12 mid-night to 12 mid-night) and pray every three hours within this, for Divine Love to reign between her and her boss. She was told to pray for all humans on Earth having difficult problems of life associated with hate, amongst others. Furthermore, I specifically directed that she should not pray for God to restore back her work, but should strictly and sincerely do as directed. These she truly obeyed.

The morning after the twenty four hours, she came to me by 6:47 am. Her aura had completely changed. Before now her aura was like a dirty white cloth, but that morning it radiated as clean as a washed white material. Then I prayed for her. She left and went straight to the head office of the bank in Lagos. People were happy to see her. The MD whose car drove in when she came was also glad to see her. She had made previous attempts to see him but could not.

The MD took this woman to his office. The solution to her problem came into manifestation, as he called the

Deeper Realities...

AGM on phone over her matter. That moment everything was resolved. The eternal potency of Love, through which the Holy Spirit works, became evident in her. The teachings of Christ came alive within her. She flew back to her base and resumed work the following day.

## ARE YOU YOUR OWN ENEMY?

In another occasion, a boy and a girl were sent by their mother from Portharcourt to see me over a family problem. The messengers reported how their earthly father was in a prolonged intense prayer every night with fasting, asking the fire of God to consume his enemy. This enemy was a certain man, a family friend, whom he suspected of having illicit sex with his wife. I sent a message of Love to him via his children, which he rebuffed whilst continuing in his vicious prayers. He even convinced his second daughter (not the one sent to me) to join him. Both of them became strong in the church wherein vicious prayers flourished.

After a long period of commanding fire, the fire certainly came, but it was his household that was consumed. He troubled his household with wicked prayers, then came the fulfillment of the Bible which has it that: "He that troubleth his own house shall inherit

Walking on the Vicious Path

the wind" (Pro. 11:29); "...and what profit hath he that laboured for the wind?" (Ecc. 5:16). Within a period of six months, his second daughter became insane, his first son was killed by armed robbers, his marriage collapsed and he went into bankruptcy.

## OVERCOMING THE PATH OF HATE

There was this woman who was having attack from the forces of darkness: She was pregnant for over three years! Doctors told her severally that it was fibroid, but she had the strong faith not to go for operation. In the course of seeking for solution she went to different 'prominent men of God' but to no avail. The bishop of her church conducted 'deliverance' yet solution did not come. A 'deliverance' pastor specifically told her that her problem was a 'senior one'.

A leading Prophet in Lagos, who told her that she was pregnant of a baby girl, ministered to her on two occasions, but solution did not come. At last she was brought to me. It was clearly made known to me that her father-in-law, a juju priest, whom she hates so much for opposing her union with his son, tied the baby in the womb with highly concentrated Astral powers. The hatred in her consciousness was the legal base of this attack, the strength from which the forces of darkness

**Deeper Realities...**

were able to accomplish this manipulation. As she got freedom from this element of hate and was transformed, prayers in Love came forth, and within twelve weeks she gave birth to a baby boy.

## CHAPTER 13

## POTENCIES OF UNIVERSAL LOVE

I remember a day when a group of fifteen people working as 'prayer warriors' met me. They were constantly engaged in what they termed 'high profile spiritual warfare against the powers of darkness'. Each of them had a tale of woe, an experience of diverse predicaments. Unknown to them, in their vicious prayers against the human agents and the forces of darkness, they drew the evidence of these into their personal lives.

I made it clear that the earthman who loves his fellow man and praises the Almighty daily in Love, lives in the Eternal Light of God; that darkness and their works do not prevail over but flee from the Light. I explained to them that such an earthman is far ahead of all those who operate with their kind of religious system. At last they went away transformed and joyful. They learnt that the best way to bind, overcome and banish the forces of darkness is not through vicious prayers, but by manifesting the potencies and Light of the Holy Spirit through Love. Darkness must flee from Light by 'Law'.

Deeper Realities...

## ASTRAL INTRICATE WEB

Therefore let the earthman refrain henceforth from commanding fire to consume his fellow man, lest the fire begin upon him by 'law'. Apart from the fire beginning on the originator, it may actually as well consume a fellow man, your enemy. Nevertheless this is not the direct work of the Holy Spirit. It is the high occult furry from a base of the Astral realm in charge of this, designed to entangle the originator in the intricate web of the vicious Astral forces. If you don't know what this stands for in your present life, you will surely know someday, with a bitter experience, in the journey of endless life.

## THE TRANSFORMING POWER

There is an important illustration from 'THE EVERLASTING GOSPEL' to the people of our world. There we read of a certain community that suffered greatly from the menace of armed robbers, kidnappers and hired assassins, all coordinated and masterminded by a notorious gang. The government tried all it could, through the police and other security agencies, to apprehend the criminals but failed.

The people of that community lived in fear day and

night, as the activities of the bad boys flourished almost on a daily basis. One evening this gang went to rob a wealthy woman who happened to be a channel of Light. They gained entrance into her house by stealth and hid themselves, waiting for the right time to strike. They could see when the woman served food on her dinning table and gathered her children to pray over the food before eating. The armed robbers, who wanted to strike as soon as the prayer ends, listened as the woman prayed.

In her prayer the woman asked God to have mercy on all men and provide food for those who don't have. She asked God to provide a means of livelihood for the teaming youths who are roaming about without job. She told God that because of this situation some of these youths, including the graduates among them, have become criminals chased up and down by the government security forces. She told God to have mercy on such criminals and transform them for good. She told God to make the government understand that, putting more effort in providing job for the youths is a better way to reduce crime than providing the security forces with more arms to fight crime. She then asked God to bless their food.

The armed robbers hidden in her house were deeply touched by the potency and the radiation of Love in this prayer. They came out. The woman and her children were terrified to see them. The gang leader told them not to fear for no harm will befall them. He told the woman that they had come to steal and kill, but that the power of her prayer has changed everything. He informed her that if she had asked God to destroy the bad boys in the community, they would have shot her dead. The vicious one told the messenger of Love that they have faith in her prayer that God would give them a better means of survival. They vowed never to rob, kidnap, or assassinate again. Even when the woman gave them money, they refused and left. They left transformed.

That was the end of all forms of kidnapping, assassination and armed robbery in that community. What the government, the police and army could not do, was done by a prayer of Love. After a while also the former bad boys were gainfully employed in different areas of work, in answer to the prayer of that channel of Light.

## LET THERE BE LOVE

The only principle that coordinates sane existence in a family, in a community, in a nation, in the entire world of man, is Love. Saint Augustine said:

> "What does Love look like? It has the hands to help others. It has the feet to

Potencies of Universal Love

hasten to the poor and needy. It has eyes to see misery and want. It has the ears to hear the sighs and sorrows of men. That is what Love looks like."

The Grail Message added:

"And what is required to grasp these principles aright? Thus to experience them? Love! And therefore Love indeed stands as the highest power, as unlimited might, in the mysteries of the great Life!".

*Prayer point*

Therefore I say to the earthman, garnish your life with the oil of Love, that you may come forth into the higher realms of Light faster in the journey of life. Always pray in Love for your fellow man. Pray in Love for the healing of the sick all over our world. Pray for the Holy Spirit to transform the prostitutes who sell themselves for money. Pray for the transformation of the armed robbers, the assassins, the kidnappers and the terrorists. Pray for the transformation of the murderers, the false Christs and false prophets, the sorcerers and all those who work for the increase of iniquity on Earth.

Also, pray for the Supreme Light to guide the path of all men on Earth; for each man to rise in Love and recognize his Oneness with God; for the Holy Spirit to direct, lead and guide the leaders of all nations for good, and much more in Love. You don't have to be a priest to

Deeper Realities...

do this. Every man is a priest in his own right. Pray with deep Love and concentration, and your prayer will go to the Almighty.

Leonard Ravenhill said:

> "The law of prayer is the law of harvest: sow sparingly in prayer, reap sparingly; sow bountifully in prayer, reap bountifully. The trouble is that we are trying to get from our efforts what was never put into them"

.Remember, you are the starting point of the answer to every prayer you make, via the ' Law of Reciprocal Action'.

## UNIVERSAL CONSTITUTION

This is not the Age to boast of being a worshipper of God in the church, in the mosque, in the temple, or on the mountain. This is the time to show Love in thoughts, words and actions. For in this Age anyone found to oppose Love will be removed from this Planet. The final consciousness that will transform humans on Earth is based on the existence of Love. The existence within this system will operate in the framework that all is good, that evil is void, and that all is well.

Therefore, let the weary ones, who are wearied in grudge against others, go now with open hearts and reconcile with their enemies. Live in Love, for this is the Age of the Holy Spirit. Always maintain a pure heart and remember the sayings of Christ:

> "Blessed are the pure in heart, for they shall see God".

Yes, they shall see God. They shall see no evil. They shall see God in action at all times through HIS Immutable Laws. Thus, they shall see God even in those who deny the existence of God. And they shall see the Manifestation and Personification of God in the realms of men.

You have been told in the course of this treatise that Earth in human terms is about to vomit. She will spew out all the elements hindering the progress of Love and Light. Surely the foremost among these are humans. To this end, let no one over struggle, fight, plot evil, hate, find fault in others, kill, maim, or become a hindrance to the progress of others, in the name of earthly survival. That chapter is closed in the Last Covenant - the current 'Universal Constitution' for all manifestations on Earth.

Let us not behave like two thieves struggling and fighting over what does not belong to them. No one

Deeper Realities...

owns anything on Earth or in the entire Universal Systems. The One Spirit of Eternity, the Holy Spirit, which is GOD, is the bona-fide Owner and LORD of all the realms of existence. If you stand firm in Love, you will manifest Light and possess what is set aside by God for you in Love within and beyond the mundane things, in the eternal journey of life. And in this, there will be no competition.

However, if you stand against Divine Love, the eternal essence of the Holy Spirit, while subjugating others in your attempt to rise in mundane wealth, power, beauty and fame, I tell you, you will soon be finished. For surely what you think you have will be taken away from you.

## WE ARE ONE

Earthman, I have told you that in the higher realms of God's Light, no one on Earth now is seen from the point of his tribe, religion, or nationality. The concept of first, second, or third world, is chaff before the Light. Each and everyone is seen from the point of the degree of Love he shows to manifest God's Divine Light in the world of man. If you fail therefore to live in Love for the manifestation of Light, then regardless of where you claim to worship God, you will not survive the winnowing of Earth even in the realms beyond.

For the portion of those who walk in the Light of Love, the Prophet Enoch of old saw the beauty of this in his time. His experience is recorded thus in 'The Lost Books of the Bible and the Forgotten Books of Eden':

> "This place, O Enoch, is prepared for the righteous, who endure all manner of offence from those that exasperate their souls, who avert their eyes from iniquity, and make righteous judgment, and give bread to the hungering, and cover the naked with clothing, and raise up the fallen, and help injured orphans... for them is prepared this place for eternal inheritance".

It is said that 'nothing goes for nothing' and that 'there is no smoke without fire'. Man does not yet truly know what he is, what the Earth is, what the Universe is and what God is. But no matter what happens, if man is ruined, then he must say to himself that he ruined himself. An

Deeper Realities...

Igbo (African) proverb has it that

> "If you allow grass to grow around your house, snakes will drink from your bathroom".

Also,

> "we should judge each day not by what we harvest but by the seed we plant".

At the level of man now in the scale of existence, he may not comprehend how what he calls 'natural disasters' are the fruits, the manifestations, of his most vicious mental act.

To transcend this, the earthman is admonished from time to time to set his consciousness to the realm of Light. Thus Saint Paul in the Bible wrote:

> "Set your affection on things above, not on things on the earth"    (Col 3:2).

The Koran (18: 104, 105) puts it this way.

> "Say, shall we tell you of those who are the greatest losers in respect of their

work? Those whose labour is all lost in search after things pertaining to the life of this world, and they think that they are doing good works"

## EVIDENCE OF LIGHT

In conclusion of this treaties, 'DEEPER REALITIES OF EXISTENCE', let it be known that it is not given to man to die first in his present life so as to enter the higher realms of Light. Whoever walks on the path of Love is in the realm of Light. The earthman in his mundane life is said to be like an 'Angel with one wing', thus he cannot fly. If he must fly, he can only do so while embracing another. (This entails the practice of Divine Love). If he must fly higher and see far, then he must first stand upon the shoulders of the tallest.

Therefore I say, in this Age service to humanity in pure Love is the evidence of Light through which you 'embrace another'. The Power of Light of the Almighty God is easily made manifest through this. Again I say, the Holy Spirit is the tallest upon whose shoulder man will stand and see far. In the past Ages, man was involved in some strenuous activities to realize God. These include seclusion in the forest or desert, going to the mountain, yoga meditation, chanting of sacred hymns, amongst others. More than all that man achieved through these are in this Age embedded in the Practice of Universal Love.

In a Divine book, Elucidations on Love, God said to man:

> " If the inhabitants of the world were to know that whoever Loves God will equally Love his brother, there would have been no war, murder, no problem whatsoever, nor the sufferings which the inhabitants of the Earth are experiencing".

Earthmen, it is said that 'when the threads unite they can tie the lion'. It is also said 'even if you choose to be an Island, you have to be friendly with the sea'. There is no deeper reality for man in creation beyond the knowledge and practice of Universal Love.

HE WHO HAS EARS TO HEAR, LET HIM HEAR.

---

**NOTE**

Questions arising from readers on the subject matter of this treatise, DEEPER REALITIES OF EXISTENCE, or on any subject related to the mysteries of life, will be published in volume two.

## BOOKS BY IYKE NATHAN UZORMA

1. OCCULT GRAND MASTER NOW IN CHRIST - Vol. 1
2. EXPOSING THE RULERS OF DARKNESS (In recognition of the highest power) Vol. 1
3. THE SPIRIT REALM Vol. 1
4. OVERCOMING THE FORCES AGAINST SUCCESSFUL MARRIAGE
5. EXPOSING THE RULERS OF DARKNESS (In recognition of the highest power) Vol. 2
6. POWERS FOR PULLING DOWN THE CONTROLLING FORCES OF DARKNESS
7. THE PATH OF PERFECTION - Vol. 1
8. THE GRAND PLAN TO DESTROY NIGERIA
9. THE OCCULTIC STRONGHOLDS IN NIGERIA AND THE REST OF THE WORLD
10. LYING SPIRIT OF THE OCCULT - Vol. 1
11. THE GREATEST REVELATION OF OUR TIME. - Vol. 1
12. WORLD-WIDE SPIRITUAL BATTLE IN NIGERIA BEFORE AND BEYOND ABACHA
13. FORMER OCCULT GRAND MASTER NOW IN CHRIST SPEAKS
14. HOW TO COMPLETELY OVERCOME WITCHES AND WIZARDS AND ALL THE POWERS OF DAKNESS
15. THE SEVEN CYCLES OF SPIRITUAL ATTACK ON BUSINESS (Exposed on earth for the first time Revised Edition)
16. TOWARDS THE REIGN OF RIGHTEOUSNESS
17. BEHOLD I GIVE UNTO YOU POWER
18. HUMAN RIGHTS ABUSE IN THE LIGHT OF REALITY

## BOOKS BY IYKE NATHAN UZORMA

19. KNOW YOUR ENEMIES WATCH YOUR FRIENDS
20. THE SPIRIT REALM Vol. 2
21. SECRETS FROM HEAVEN
22. THE GREATEST EVIDENCE OF GOD'S POWER IN OUR TIME
23. OCCULT GRAND MASTER NOW IN CHRIST - Vol. 2
24. OCCULT GRAND MASTER NOW IN CHRIST - Vol. 3
25. PARADOX OF EXISTENCE
26. THE GREATEST STRANGE BEING NOW ON EARTH
27. THE FUTURE EARTH
28. MY 300 MINUTES EXPERIENCE OF HEAVEN
29. THE BOOK OF TESTIMONIES - Vol. 1
30. THOU SHALL DECREE A THING
31. THE BOOK OF LIGHT
32. DEEPER REALITIES OF EXISTENCE Vol.1
33. HIDDEN TRUTH OF MAN AND WOMAN
34. THE PATH PERFECTION Vol.2
35. EARTHMEN RETURN TO THYSELF
36. THE KINGDOM OF GOD IS WITHIN YOU

CPSIA information can be obtained
at www.ICGtesting.com
Printed in the USA
BVHW081918210220
572978BV00003B/298